Climate Change and Energy Insecurity

Climate Change and Energy Insecurity

The Challenge for Peace, Security and Development

Edited by

Felix Dodds, Andrew Higham and Richard Sherman
with a foreword by Achim Steiner

Routledge
Taylor & Francis Group
LONDON AND NEW YORK

earthscan
from Routledge

First published 2009 by Earthscan

Published 2022 by Routledge
2 Park Square, Milton Park, Abingdon, Oxon OX14 4RN
605 Third Avenue, New York, NY 10017

Routledge is an imprint of the Taylor & Francis Group, an informa business

Publisher's Note

The publisher has gone to great lengths to ensure the quality of this reprint but points out that some imperfections in the original copies may be apparent.

ISBN: 978-1-844-07855-4 (hbk)
ISBN: 978-1-84407-856-1 (pbk)

Typeset by JS Typesetting Ltd, Porthcawl, Mid Glamorgan
Cover design by John Charles

A catalogue record for this book is available from the British Library

Library of Congress Cataloging-in-Publication Data
Climate change and energy insecurity : the challenge for peace, security, and development / edited by Felix Dodds, Andrew Higham and Richard Sherman; with a foreword by Achim Steiner.
p. cm.
Includes bibliographical references and index.
ISBN 978-1-84407-855-4 (hardback) – ISBN 978-1-84407-856-1 (pbk.) 1. Climatic changes–Political aspects. 2. Energy development–International cooperation. 3. Security, International. I. Dodds, Felix. II. Higham, Andrew. III. Sherman, Richard.
QC903.C544 2009
363.7'0526--dc22

2009027987

Contents

List of Figures, Tables and Boxes

Figures

Tables

Boxes

List of Contributors

Linda Adams is secretary of the California Environmental Protection Agency (Cal/EPA). Formerly director of the California Department of Water Resources, she was appointed by Governor Arnold Schwarzenegger in May 2006, making her the first woman to serve as head of the agency. As Cal/EPA's secretary, Adams oversees the environmental activities of the Air Resources Board, California Integrated Waste Management Board, State Water Resources Control Board, Office of Environmental Health Hazard Assessment, Department of Toxic Substances Control, Department of Pesticide Regulation, and the approximately 5000 employees who serve the state's diverse environmental programmes. She is also Governor Schwarzenegger's lead negotiator on AB 32, the Global Warming Solutions Act of 2006, and in this capacity works closely with states, provinces and countries around the world to develop a network of climate initiatives. As secretary, Adams has been instrumental in a number of environmental policies, including launching the Green Chemistry Initiative, which calls for a more comprehensive look at California's chemical policies, accelerating air quality improvement, and the state's bio-monitoring programme, which seeks to link environmental contaminants to human health. Adams has previously served in Governor Davis's administration as director of the Department of Water Resources and as chief deputy legislative secretary. Adams's most notable accomplishments include the Clean Cars/Clean Air legislation, earning her the Environmental Hero Award from the California League of Conservation Voters. Prior to this, she served as a staff member to the State Legislature for 20 years in various positions. She currently serves on the board of directors of the Sacramento Local Conservation Corps and is a long-time resident of Sacramento, California.

Nnimmo Bassey is a Nigerian human/environmental rights activist. He is the executive director of Environmental Rights Action (ERA) – Nigeria's foremost environmental rights advocacy group. He is also chair of Friends of the Earth International – the world's largest federation of grassroots organizations fighting for environmental and social justice. Friends of the Earth groups are found in 77 countries. Bassey is a member of the international steering committee of Oilwatch International. He is also a practising architect in Nigeria as well as a published writer and poet.

Ambassador Stuart Beck is permanent representative of the Permanent Mission of the Solomon Islands to the United Nations since 2004. Ambassador Beck holds an MA in international relations and South-East Asian politics from the University of Queensland, a foreign services certificate from Oxford University, and a BA in administration and political science from the University of the South Pacific. Ambassador Beck is an avid spokesman for small island states and the severe threats that they face due to climate change.

Craig Bennett is a deputy director at the University of Cambridge Programme for Sustainability Leadership (CPSL) and, in this role, is also co-director of The Prince of Wales's Corporate Leaders Group on Climate Change (CLG). In late 2007, he led the CLG's work on the Bali Communiqué, which brought together 150 of the world's largest companies in support of a comprehensive, legally binding United Nations framework to tackle climate change. A year later, he led the work that resulted in the Poznan Communiqué in which companies from Europe, the US, Brazil, India, Japan, China and Australia agreed on some of the key elements needed for a global climate deal. Both initiatives generated significant political interest and media attention. Prior to joining CPSL, Bennett was the head of the Corporates, Trade and Globalization Campaign at Friends of the Earth. He also sat on the executive committee of Friends of the Earth International, the steering group of the Corporate Responsibility Coalition (CORE) and was a board member of the Trade Justice Movement (TJM). He has a BSc (Hons) in Human and Physical Geography and an MSc in Biodiversity Conservation. He is a Fellow of the Royal Geographical Society (FRGS) and a Fellow of the Royal Society of Arts (FRSA). He is also a visiting lecturer at the University of Bristol and the London Business School, sits on the board of Stakeholder Forum (the United Nations Environment Programme (UNEP) National Committee for the UK) and on the editorial board of *Sustainable Business* magazine. He is a member of the Corporate Responsibility Advisory Panel for the retail group Kingfisher.

Barbara J. Bramble is senior adviser for international affairs at the National Wildlife Federation (NWF) and a strong advocate to improve US policy regarding climate change, tropical deforestation and unsustainable agriculture production and trade. Over two decades, as the original founder and director of NWF's International Affairs Department, she developed innovative coalitions and grassroots support to reform the World Bank and other multilateral development banks; promote US leadership to reduce the threat of climate change; help to establish voluntary certification systems to promote sustainable forest and agriculture products; and advocate that international trade agreements promote, rather than frustrate, sustainable development. Bramble was a key organizer of the 1992 Earth Summit in Rio de Janeiro and the Rio +5 Conference in 1997. More recently, for four years she worked with Mexican non-governmental organizations (NGOs) to enhance their advocacy and environmental education skills. In her current role as NWF's senior adviser, she is helping to develop international sustainability standards to guide the burgeoning biofuels industry, serving as chair of the steering board of the Roundtable on Sustainable Biofuels; she is also the immediate past chair of the Forest Stewardship Council in the US. She serves on the boards of several non-profit organizations in Mexico, Brazil and the US. She is an environmental lawyer, and has worked in the Executive Office of the US president, and in private law practice.

Gidon Bromberg is the Israeli director of EcoPeace/Friends of the Earth Middle East. Friends of the Earth Middle East is a unique regional organization that brings together Jordanian, Palestinian and Israeli environmentalists to promote sustainable development and advance peace efforts in the troubled Middle East. The organization has offices in Tel Aviv, Bethlehem and Amman, employs 37 paid staff and actively involves hundreds of volunteers. Bromberg founded the organization under the name of EcoPeace in 1994 and has been the Israeli director ever since. In 1997 he led the entry of the organization into Friends of the Earth International. He has spearheaded the advocacy campaigns of the organization both in Israel and internationally, and developed the cross-border community peace-building programme Good Water Neighbours that is seen as a model for other programmes in conflict areas. Bromberg speaks regularly on water, peace and security issues; including at the United Nations Commission for Sustainable Development; before the US House of Representatives, International Relations Committee; the European Parliament; and before the advisory meeting to the United Nations High-Level Panel on Security. In 2007 he was invited to join the prestigious East West Institute's International Task Force for Preventive Diplomacy. He was most recently selected for the 2007 World Fellowship at Yale University on global leadership. Bromberg is a member of the Israeli inter-ministerial committee on the future of the Dead Sea, of the Israel United Nations Educational, Scientific and Cultural Organization (UNESCO) World Heritage Committee and the inter-ministerial Committee for Sustainable Development in Israel. He has published over 20 academic and popular publications concerning Middle East environmental policy and water security issues.

Stavros Dimas is a Greek politician and, currently, European Commissioner for the Environment. He worked as a lawyer for the World Bank and Sullivan & Cromwell, a Wall Street law firm. Commissioner Dimas has been elected to the Greek parliament ten consecutive times since 1977 and has held leading posts in Greek politics. From 2000 to 2003 he was senior member of the Political Analysis Steering Committee of New Democracy and in 2004 he was briefly European commissioner for employment and social affairs. Commissioner Dimas has studied law and economics in Athens, Greece, and has an MA in Law from New York University. He considers climate change, preserving biodiversity and fostering eco-innovation to be key priorities for the European Union.

Jacques Diouf is director-general of the United Nations Food and Agriculture Organization (FAO). Dr Diouf has held several positions of responsibility in national and international agricultural institutions: director of the European Office and the Agricultural Programme of the Marketing Board (Paris/Dakar); executive secretary of the African Groundnut Council in Nigeria; executive secretary of the West Africa Rice Development Association in Liberia; and adviser to the president and regional director of the International Development Research Centre, Ottawa (Canada). His expertise was put at the service of the executive and legislative branches of the Senegalese government as secretary of state for science and technology in the Cabinet and member of parliament; chairperson of the Foreign Relations Committee and secretary-elect, Dakar (Senegal); and ambassador, Permanent Mission of the Republic of Senegal to the United Nations, New York. He also served as secretary-general and special adviser to the governor of the Central Bank for West African States, Dakar (Senegal). He sat on the boards of a number of international institutions and produced several scientific publications. He has received numerous high awards from governments around the world and various academic recognitions (Doctor Honoris Causa and Membership of Academies of Agriculture and Science).

Ahmed Djoghlaf is an Algerian national. He has pursued a distinguished diplomatic career that has included postings with the government of Algeria and UNEP. He assumed the position of executive secretary of the Convention on Biological Diversity (CBD) on 3 January 2006. His previous positions included assistant executive director of UNEP and director and coordinator of UNEP's division of the Global Environment Facility (GEF). He was also the general rapporteur of the Preparatory Committee of the United Nations Conference on Environment and Development (UNCED) (the Rio Earth Summit), vice chairman of the Eleventh Session of the Intergovernmental Committee on Science and Technology for Development, and vice president of the negotiating committee on the Framework Convention on Climate Change. Dr Djoghlaf has also chaired one of the two negotiating committees of the Convention to Combat Desertification (CCD). Prior to joining the United Nations, Dr Djoghlaf held a variety of key posts in the Algerian Ministry of Foreign Affairs. He was adviser on environmental issues to the prime minister of Algeria and, prior to that, to three ministers of foreign affairs of Algeria. He holds the rank of minister plenipotentiary of the Ministry of Foreign Affairs and assistant secretary-general of the United Nations.

Felix Dodds is the executive director of Stakeholder Forum for a Sustainable Future (SF). He has been active at the United Nations since 1990, attending the World Summits of Rio, Habitat II, Rio +5, Beijing +5, Copenhagen +5, the World Summit on Sustainable Development (WSSD) and World Summit 2005. He has also been to all of the United Nations Commissions for Sustainable Development and the UNEP Governing Council meetings. He has set up three global NGO coalitions for UN conferences, summits and commissions: the

United Nations Commission on Sustainable Development (1993); UN-Habitat II (1995); and the World Health Organization (WHO) Health and Environment Conference (1999). He co-chaired the NGO coalition at the United Nations Commission on Sustainable Development from 1997 to 2001. He introduced stakeholder dialogues in 1996 through the United Nations General Assembly in November 1996 for Rio +5 and helped to run some of the most successful ones at Bonn Water (2001) and Bonn Energy (2004). He has written or co-edited the following books: *Negotiating and Implementing Multilateral Environment Agreements* (UNEP, 2007); *Human and Environmental Security: An Agenda for Change* (Earthscan, 2005); *How to Lobby at Intergovernmental Meetings: Mine is a Café Latté* (Earthscan, 2004); *Earth Summit 2002: A New Deal* (Earthscan, 2000); *Multi-Stakeholder Process on Governance and Sustainability: The Way Forward – Beyond Agenda 21* (Earthscan, 1997); and *Into the Twenty-First Century: An Agenda for Political Realignment* (Green Print, 1988). He occasionally writes for the BBC Green Room and blogs from film festivals.

Michael K. Dorsey is assistant professor in Dartmouth College's Environmental Studies Programme, UK. He is a co-contributor of the recently re-released new edition volume *Climate Change, Carbon Trading and Civil Society: Negative Returns on South African Investments* (Rozenberg Press, The Netherlands, and University of KwaZulu-Natal Press, South Africa, 2008). For more than a decade Dr Dorsey has provided strategic guidance and advice to governments, foundations, firms and a multitude of others on the interplay of multilateral environment policy, finance and economic development matters. In 1992, he was a member of the US State Department Delegation to the United Nations Conference on Environment and Development (Rio Earth Summit). From 1994 to 1996 he was a task force member of President William Jefferson Clinton's Council on Sustainable Development. A member of the Sierra Club (www.sierraclub.org) since the mid 1980s, Michael served six years (1997 to 2003) on the club's national board. In 1997, in Glasgow, Scotland, Dr Dorsey was bestowed Rotary International's highest honour: the Paul Harris Medal for Distinguished Service to Humanity. Fluent in Spanish, from 1999 to 2001, Dr Dorsey lived in Ecuador and was a programme director based jointly with the Instituto de Estudios Ecologistas del Tercer Mundo and Acción Ecológica. While at Acción Ecológica Dr Dorsey co-led an initiative to monitor the commercialization of biodiversity in Ecuador as well as the other four countries in the Andean Pact. He is a graduate of the University of Michigan School of Natural Resources and Environment (BSc and PhD), Yale University's School of Forestry and Environmental Studies (MFS) and The Johns Hopkins University (MA). Before joining Dartmouth's faculty, he held the college's Thurgood Marshall Fellowship in Environmental Studies and Geography. Over the years, Dr Dorsey has also held visiting positions at various institutions around the world, including the Rijksuniversiteit Groningen at the Royal University of Groningen (The Netherlands) in 2001; the Department of Regional Planning at the Kungliga Tekniska Högskolan (Royal Institute of Technology, Sweden) in

2002; and the Graduate School of Public and Development Management at the University of Witswatersrand (South Africa) in 2002. From April 2007 until November 2008 Dr Dorsey served as a member of Senator Barack Obama's Energy and Environment Presidential Campaign team. In January 2009 the Ford Foundation recognized Dr Dorsey's ongoing contribution to examining the interplay of finance, environmental policy and social justice concerns with a grant to support his research on how emerging carbon markets shape justice-based climate policies.

Elizabeth Dowdeswell was, until recently, president of Canada's Nuclear Waste Management Organization (NWMO). NWMO's initial purpose was to conduct a study and develop a recommendation for the Government of Canada on the long-term management of Canada's used nuclear fuel. Having completed this assignment, she continues as special adviser to the board of NWMO. Dowdeswell has served as executive director of UNEP and undersecretary general of the United Nations, and assistant deputy minister of Environment Canada, responsible for the national weather and atmospheric agency, including negotiating the Framework Convention on Climate Change. She has also led a number of public inquiries into such politically sensitive issues as Canada's unemployment benefits programme and federal water policy. Her early career included terms as deputy minister of culture and youth for the Province of Saskatchewan, educational consultant, university lecturer and high-school teacher. She was appointed one of the first Pierre Elliot Trudeau Foundation mentors, helping to guide the public policy research of Trudeau scholars. She is also a visiting professor in genomics and global health at the Joint Centre for Bioethics, University of Toronto. Dowdeswell is a director on the board of several corporations and contributes in an advisory capacity to a number of not-for-profit organizations, including chairing the Scientific Advisory Committee of the new Council of Canadian Academies.

Ladeene Freimuth served as the deputy director of Friends of the Earth Middle East, where she worked on climate change and transboundary water management issues. She is also the founder of a domestic and international energy and environmental consulting firm, the Freimuth Group, LLC. She specializes in policy and strategic advising, lobbying, research and analysis, and project design, implementation and management. Her areas of expertise include renewable energy, energy efficiency, global climate change, alternative fuels and transportation, transboundary water management, and sustainable development. She has worked for 17 years with the US Environmental Protection Agency and the US Agency for International Development, a number of consulting firms and not-for-profit organizations. While working in US Congress, she drafted the first piece of legislation to reduce air and greenhouse gas emissions from stationary power plants. She has also has written major pieces of energy efficiency, renewable energy and alternative transportation legislation. Freimuth received a BA in Government from Wesleyan University and an MA in International Resource Policy and Eurasian Studies from the

Fletcher School of Law and Diplomacy at Tufts University, US, with emphases on global warming and transboundary water management.

Sigmar Gabriel is currently serving as federal minister for the environment and nature conservation and nuclear safety minister in the cabinet of Angela Merkel, Germany. He is a member of German Social Democratic Party. In 1999 he was the minister-president of Lower Saxony and served until 4 March 2003. He held other posts in the Lower Saxony government. In 2008 he acted as president of the Ninth Conference of the Parties to the Convention on Biological Diversity.

Winston Gereluk represents Public Services International (PSI) on matters relating to the Commission on Sustainable Development (CSD) and environmental issues, generally. The PSI is an international trade secretariat for public service trade unions, and as such represents millions of public employees around the world. Gereluk has participated in delegations and the production of literature on behalf of the Global Unions at the United Nations CSD and is a member of Greenpeace. He is employed by the Alberta Union of Provisional Employees, an affiliate of the National Union of Public and General Employees in Canada. He also instructs and develops industrial relations courses for Athabasca University in Alberta, Canada.

Devyani Gupta has worked on sustainable development, humanitarian aid, refugees and migration with a number of organizations, including the European Commission (EC), the United Nations High Commissioner for Refugees (UNHCR), Stakeholder Forum for a Sustainable Future and the International Institute for Environment and Development. She is fluent in Hindi and holds an MA (University of Chicago) in International Relations. She is currently working as a consultant, advising on health and social care issues with assignments at the UK Department of Health, National Health Service (NHS) organizations and local government.

Andrew Higham is currently Issue Manager supporting global climate change negotiations on technology development and transfer within the United Nations Framework Convention on Climate Change. He was previously an expert with the International Climate Policy Group within the Energy Research Centre of the Netherlands (ECN) where he advised the UNFCCC on finance issues, including the development of the EGTT recommendations on financing options to the AWG-LCA, and on financing indicators for monitoring the implementation of the Convention. He was also a leading Australian adviser in sustainability, climate change, marine and water policy. He co-authored the Western Australian State Sustainability Strategy and the 1998 *State of the Environment Report*, and has led or contributed to a wide range of government policies, strategies, task forces and reports. He has served as the strategies director and vice president of the Australian Conservation Foundation, a ministerial adviser in the fields of science, environment, water and climate change, and has played a leading role in a several Australian environmental policy reforms, including as a member of the Australian Emissions Trading Taskforce. He was previously an expert advisor to the United Nations Division of Technology, Industry and Economics (DTIE) International Environmental Technology Centre, and helped to establish the Network of Regional Governments for Sustainable Development at the World Summit on Sustainable Development in 2002.

Steve Howard is the chief executive officer of The Climate Group, an organization that he co-founded in 2003 to bring together business and government and set the world economy on the path to a low-carbon prosperous future. Howard has worked on global social and environmental issues from within business, NGO and UN settings. He has lead the establishment of the Climate Group in the US, Australia, China, India, Canada and Brussels. He chairs the World Economic Forum's Global Agenda Council on Climate Change and is an adviser to the Virgin Earth Challenge. Prior to his role with The Climate Group, Howard was a partner in the consultancy ERM, where he led on corporate social responsibility. Previously he was the founding director of the Global Forest and Trade Network for the World Wide Fund for Nature (WWF) International and chairman of the 95+ Group. He was also chairman of the UK Forest Stewardship Council, where he helped to achieve a far-reaching consensus on a national forest certification standard, and founding chairman of the Tropical Forest Trust, which he helped establish to deliver ethically sourced tropical timber.

Nader al Khateeb holds a BSc in Geological Engineering from the Middle East Technical University in Turkey and an MSc in Environmental Management from the Loughborough University of Technology in the UK. He is the general director of the Water and Environmental Development Organization (WEDO), a Palestinian organization active in the environmental sector. He is also the Palestinian director of Friends of the Earth Middle East, a regional organization that brings together Jordanian, Palestinian and Israeli environmentalists to promote sustainable development of the shared resources in the Middle East. He has over 20 years of experience in water management. Khateeb was a senior water resource engineer with the United Nations Development Programme (UNDP), working as a consultant on the ultimate creation of the Palestinian Water Authority (PWA), and was then a senior water resource engineer with the PWA, coordinating water and wastewater projects. He carried out feasibility studies of the Industrial Waste Management in Hebron, worked with UNESCO to prepare a conceptual Emergency Master Plan for the Bethlehem region and trained municipal engineers in the Gaza Strip on wastewater collection and treatment systems. Khateeb also served as chief engineer for the Bethlehem, Beit Jala and Beit Sahour Water Authority and project manager for their drainage and sewerage project.

Melinda Kimble is a senior vice president at the United Nations Foundation and oversees its International Bioenergy Initiative. She joined the United Nations

Foundation in May 2000; prior to this, she served as a state department foreign service officer, attaining the rank of minister-counsellor. She served in policy-level positions in the Bureau of Economic and Business Affairs, overseeing multilateral development issues and debt policy; in the Bureau of Oceans, International Environment and Scientific Affairs (OES), leading environmental negotiations (e.g. the Climate Change Conference in Kyoto, Japan, in 1997). Her assignments abroad include the Côte d'Ivoire, Egypt and Tunisia. She speaks French and Arabic and holds two Master's degrees: Economics (University of Denver) and MPA (Harvard's Kennedy School of Government). In a personal capacity she serves on the Board of International Science Organizations of the National Academies of Science and on the US National Commission for UNESCO. She is a member of the Women's Foreign Policy Group and the Women's Council on Energy and Environment.

Johan Kuylenstierna currently works as the chief technical adviser to the chair of United Nations-Water, an interagency mechanism supporting increased collaboration among the 26 United Nations agencies and programmes with water issues as part of their mandate. He is based at the FAO in Rome. He also holds an adjunct professorship at Stockholm University. Johan has previously worked as a project director at the Stockholm International Water Institute, as a senior consultant focusing on corporate social responsibility, corporate core value processes and international stakeholder consultation processes. He also has previous experience from the United Nations system as a scientific officer at the World Meteorological Organization and as a junior professional officer at the Division for Sustainable Development (UNDSD) at the United Nations Headquarters in New York. He has worked for many years as a university teacher and has his research background in palaeoclimatology and physical geography.

Jürgen Maier has been the director of the German NGO Forum Environment and Development in Berlin since 1996. This network coordinates the participation of Germany's NGOs in international negotiations on sustainable development, such as the United Nations Framework Convention on Climate Change (UNFCCC), the Convention on Biological Diversity, World Trade Organization (WTO) summits, etc. He is a board member of the Roundtable on Sustainable Biofuels (RSB). Maier was formerly the director of the German Asia Foundation during 1993 to 1996 and international secretary of the Green Party Germany during 1987 to 1991.

Munqeth Mehyar received his BA in Regional Planning and Architecture from the University of Louisiana, US, in 1981. In 1994, he co-founded the regional environmental organization EcoPeace/Friends of the Earth Middle East (FoEME). He is chairman of the organization and director of the Amman office, leading FoEME activities concerning the Jordan River, Dead Sea and the Gulf of Aqaba. As chairman, his responsibilities include the supervision of international project development and management; liaison and lobbying of governmental and private-sector figures on major regional policy issues relevant to environmental protection; development of international contacts and functional partnerships with international environmental and development institutions; and has represented the organization in international and regional forums and conferences. He has also organized dozens of regional conferences, workshops and study tours, has supervised regional research on shared ecosystems and co-authored reports and policy papers. He is on the board of directors of both the Jordan Sports Federation for the Handicapped and the Jordan Royal Ecological Diving Society. He is also founder and director of the Jordan Society for Sustainable Development, an environmental NGO based in Amman. Recently, he received a medal from Al Houssain Distinguished Giving.

Michael Moore is currently a project manager at the Stockholm International Water Institute (SIWI), where he has worked since 2005. His main responsibility within SIWI relates to programme coordination of the World Water Week, an annual conference in Stockholm that brings together water experts and professionals around issues related to water and development. Moore also works in the Swedish Water House, specializing in environmental flows, agricultural water management and freshwater ecosystems. He has been involved in the Comprehensive Assessment of Water Management in Agriculture and is currently a steering committee member for the Global Environmental Flows Network. Prior to SIWI, he was based in Brisbane, Australia, as an environmental project management consultant. His technical background is in ecology and land and water processes, with a focus on the riverine systems in western Queensland, Australia.

Derek Osborn is the president of Stakeholder Forum. He is also a member of the European Economic and Social Committee and vice president of their Sustainable Development Observatory. Previously, he was director general of the UK Department of the Environment, chairman of the European Environment Agency, and chair or member of several other bodies concerned with sustainable development in the public, private and voluntary sectors. In 1997 he co-chaired the United Nations General Assembly negotiations assessing global progress on sustainable development five years after the Earth Summit in Rio, producing a programme for its further implementation. His *book Earth Summit II: Outcomes and Analysis*, co-authored with Tom Bigg and with a foreword by The Right Hononourable Tony Blair, was published by Earthscan in 1998.

Lucien Royer is director of the Canadian Labour Congress (CLC) International Department. Until July 2008 he was the health, safety and environment officer for both the Global Unions and the Trade Union Advisory Committee (TUAC) to the Organisation for Economic Co-operation and Development (OECD) where he coordinated programmes throughout the world and represented trade unions at the United Nations, OECD and other intergovernmental bodies with respect to health, safety and environment. He currently chairs UNEP's Major Group Facilitating Committee. Originally from western Canada, he was involved in environmental litigation and environmental law reform and was a founding member of the Canadian Environmental Network. He formed part of the Canadian government delegation at the European preparatory meeting for the Earth Summit in 1990 and at Rio in 1992. He has led the trade union delegation at UN CSD meetings since 1993.

Marthinus van Schalkwyk is minister of environmental affairs and tourism of the Republic of South Africa. He has been a member of parliament since 1990, and from 2002 to 2004 was the premier of Western Cape Province. He lectures in political science at the Rand Afrikaans University and University of Stellenbosch, South Africa. Van Schalkwyk has led the South African delegation to international climate change negotiations since 2004 and currently holds the chairpersonship of the African Ministers Conference on Environment (AMCEN). He is also director of the Business Trust, an initiative that combines the resources of the government and business to create jobs, build capacity and combat poverty.

Richard Sherman serves as the programme manager for the International Institute for Sustainable Development Reporting Services' (IISD RS) Africa Regional Coverage Project. He was a former team leader and writer from 2002 to 2005. In addition, Sherman works as a consultant to Stakeholder Forum for a Sustainable Future and the South African Department of Environmental Affairs and Tourism, with a particular focus on institutional reform and global governance. From 1998 to 2001 he was a member of the South African government's climate change delegation and has played an active role in the South African NGO sector since 1995. His previous books were *10 Days in Johannesburg: A Negotiation of Hope*, with Pam Chasek and Chris Spence (UNDP and South Africa's Department of Environmental Affairs and Tourism, 2004); and *Environment and Development Decision Making in Africa 2006–2008* (IISD, 2008). He previously worked for GLOBE and Earthlife Africa, as well as being an active member of the Climate Action Network (CAN).

Andrew Simms is currently head of the global economy programme at the think tank the New Economics Foundation (nef) and an adviser to Christian Aid, the overseas development agency. In the past he studied at the London School of Economics and worked in politics for a range of development and environment organizations. In particular, he has been part of campaigns on debt, climate change, food security and genetically modified food, and corporate accountability. He is currently working on a Global Green New Deal.

Chris Spence is the deputy director of reporting services at the International Institute for Sustainable Development (IISD) and manages the Climate Change Team for the institute's well-known publication, *Earth Negotiations Bulletin* (*ENB*). He has written and lectured on climate change for many years and is the author of *Global Warming: Personal Solutions for a Healthy Planet* (Palgrave Macmillan, 2005).

Nicholas Stern (Lord Stern of Brentford, Kt, FBA) is IG Patel Professor of Economics and Government at the London School of Economics (LSE), where he is also head of the India Observatory within the LSE's Asia Research Centre and chairman of the Grantham Research Institute on Climate Change and the Environment. In 2005, he was appointed by the UK government to conduct the influential Stern Review on the Economics of Climate Change. Prior to that he was head of the Government Economic Service and, from 2003 to 2005, Second Permanent Secretary to Her Majesty's Treasury; from 2004 to 2005 he was director of policy and research for the Prime Minister's Commission for Africa. Previously, having held academic posts at the universities of Oxford and Warwick and the LSE, he was then chief economist for the European Bank for Reconstruction and Development and, subsequently, chief economist and senior vice president of the World Bank. Research and publications have focused on the economics of climate change, economic development and growth, economic theory, tax reform, public policy, and the role of the state and economies in transition. He has published more than 15 books and 100 articles. He has served on committees for Oxfam, the Overseas Development Agency (ODA) and the United Nations. He is a fellow of the British Academy (since July 1993) and has Foreign Honorary Membership of the American Academy of Arts and Sciences (since 1998) as well as honorary doctorates from a number of universities, including Warwick and Cambridge.

Dave Trouba is head of communications at the Water Supply and Sanitation Collaborative Council (WSSCC) in Geneva, Switzerland. Prior to joining WSSCC in 2008, he worked for ten years at the Stockholm International Water Institute (SIWI), serving as its communications director. He was responsible for the overall communications, marketing and media strategies for SIWI and its various sub-programmes, including the World Water Week in Stockholm. From 1991 to 1998, he was the public information manager at the Water Environment Federation (WEF) in Alexandria, Virginia, US. Upon completing his degree in journalism from Benedictine College in Atchison, Kansas, US, he worked with Communications and Conference Coordination at JT&A, an environmental consulting firm in Washington, DC. His professional philosophy is based on the belief that communications play an important role in bridging the divide between knowledge and action in our efforts to overcome water, sanitation, environment and human development challenges. In his spare time, he co-hosts the radio programme *Think Globally* (www.thinkgloballyradio.org).

Foreword

Achim Steiner, UNEP Executive Director

The race to mitigate climate change is not just about preserving the environment, it's also about security. The security in knowing that our homes are safe, that we have enough food to eat, and that our water is drinkable. The security in knowing that crops will grow, and that there will still be fish in the sea. The security also that comes with caring for our neighbours – and knowing that they too have food to eat. Because our changing climate is changing more than just the weather, it is changing the way we live.

Inhabitants of small islands are watching their coastlines disappear. Weather changes are affecting plant life, and the diets of the animals and humans who eat those plants. Diminishing water reserves are creating strife between and within countries, causing some populations to migrate. Clearly, relatively minor changes in global temperature have major implications in many aspects of our lives.

Much of the pollution that is behind the causes of climate change is related either directly or indirectly to energy use. But we can't just turn off the switch – energy access is essential to growth and prosperity. If modern society is dependent on abundant energy sources, and if using those sources threatens the planet's future, what kind of future are we looking at? Renewable and sustainable energy resources are not only essential to stabilizing global climate, but also to securing lasting energy resources and energy uses for the millennia to come.

Among other issues, certain energy sources, like fossil fuels, may soon run out. Our dependence on fossil fuels is not only affecting our atmosphere, but it is also destabilizing entire regions. As oil reserves dwindle, competition increases, as does the threat of price wars, and even real wars, over what's left. Even under the most optimistic scenario, the race to find new oil reserves will be a short one. Energy security will become an increasingly important issue, both for the planet and for our own well-being.

This timely publication addresses the multifaceted character of climate change, where each separate element affects other elements, much like the environment itself. Each chapter looks at climate change and energy insecurity from a different angle, be it energy options, economic changes, biodiversity issues, or many others. It also addresses an important question: how are we going to clean up the mess? What kind of international agreement can be achieved? What kind of frameworks work? Hopefully, the answers to at least some of these questions will be found in Copenhagen in December 2009 as governments meet to seal a new climate deal. This is our chance to move forward towards ensuring that both our climate and our energy sources are as secure as the ground beneath our feet.

Achim Steiner
United Nations Under Secretary General
United National Environment Programme Executive Director

Preface: Tomorrow Is Today

Felix Dodds, Andrew Higham and Richard Sherman

Reality is merely an illusion, albeit a very persistent one.
(Albert Einstein)

Increasingly, climate change policy is being recognized and addressed across the entire multilateral framework. It is no longer the sole province of the United Nations Framework Convention on Climate Change (UNFCCC); yet there remains a vast amount of work to fully mainstream climate change within the multilateral system. Indeed, the peace and security implications of climate change are only just beginning to be acknowledged. This book started out as a contribution to the preparations for the Poznan UNFCCC meeting in 2008. As the authors took stock of the landscape, they felt that it was more important to be seen outside of the UNFCCC meetings. The issues raised in this book are broader than those that are going to be addressed in the climate change negotiations.

The challenges

Climate change is recognized as one of the most important challenges facing the international community. Many eminent politicians have declared climate change as the biggest threat to our civilization; others have suggested that global warming is more of a threat than the traditionally viewed risks to international peace and security. The science is compelling and many of the leading climate change scientists are concerned about abrupt and potentially dramatic climate and other Earth system tipping points, with potentially massive impacts upon life on Earth over centuries and geological time frames that could become irreversible without dramatic technological and behavioural changes beginning now.

In 2007, for the first time, an environmental issue reached the United Nations Security Council. This discussion represented a realization of the political impact of not implementing the agreements at the Rio and Johannesburg Earth Summits. Within the next 10 to 20 years, environment and, in particular, the climate implications are going to move to become security issues. The authors have addressed both the problems of accelerated energy needs and the impacts that climate change could have upon peace, security and development. The book has also tried to address the global institutional framework that is needed to address the challenges.

Part I – Energy Insecurity: Challenges to Future Energy Stability

In this section of the book authors present various analyses on the challenges facing the international community in meeting future energy demands by highlighting the potential security-related problems that may arise from technological choices, energy uses and, in some cases, land-use choices. The section will aim to contextualize the problems related to both traditional and new energy resources by highlighting areas where conflicts, either localized or global, may arise in the future. This includes a focus on security issues related to the oil and gas, nuclear, biofuels and hydropower sectors, highlighting areas that may cause potential threats to international peace and security, as well as to development. This is not all the challenges we are facing; the section tries also to look at energy access for all and the meeting of the Millennium Development Goals, addressing sustainable consumption and production in the rapidly industrializing development countries, as well as the challenge of moving away from present consumption patterns in industrialized countries to rapidly sustainable consumption patterns.

Part II – Climate Insecurity: A Challenge to Peace, Security and Development

This section of the book presents an overview of the potential security threats and challenges posed by climate change. It addresses a number of key sectoral issues where climate change will lead to elevated concerns to peace, security and development.

Part III – Governing Climate and Energy Instability: Avenues for Preventative Diplomacy

Finally, this section of the book explores the potential avenues and mechanisms available to the international community to address and avert climate and energy instability via the multilateral framework under the United Nations. The chapters in this section address a number of forward-looking options for international governance, and focus on the appropriate forums to discuss and agree on preventative approaches within the current climate and energy frameworks, such as the UNFCCC, the United Nations General Assembly and the Security Council. The section also addresses appropriate mechanisms for the transfer of resources from industrialized to developing countries to avoid the impacts of climate change and to support a sustainable transition away from climate-polluting development. It also provides an analysis on the steps needed to ensure an energy transition by focusing on the rapid deployment of clean energy technologies and how to address job insecurity created by the transition away from traditional energy sources. The section concludes by outlining the existing avenues where climate and energy insecurity issues can be addressed.

And then

This book is a companion to *Human and Environment Security: An Agenda for Change*, edited by Felix Dodds and Tim Pippard (Earthscan, 2005).

We believe that we are the first generation who realizes that our personal actions affect, but also contribute to, the sustainability of our planet. We have, perhaps, a short time: 10 to 20 years to redirect the development model that has dominated this period of our lives on this planet. We know that it is destroying options for future generations. On this generation falls the burden of proving to the world that we really mean it when we say that all people are created free and equal and should benefit from the fruits of this planet to ensure that we live in a sustainable way.

We might all wish for an easier challenge, a more tranquil world; but that isn't our lot. If we are to create a sustainable world for all living people and future generations to come, then we must change now, not in 10 to 20 years. Let us hope that Einstein was not right when he said: 'Two things are infinite: the universe and human stupidity; and I'm not sure about the universe.'

Special thanks

This book would not have been possible without the support of a number of key people whom we would like to give special thanks to: Rob West, Claire Lamont, Jonathan Sinclair Wilson and Sarah Thorowgood at Earthscan; Emily Benson, Ida Bergstrøm, Lianna Hulbert, Sonia Misquitta and Stephen Mooney, from Stakeholder Forum; Panagiota Stylianou and Maria Gerolymatou at the European Commission; Luc Bas from the Climate Group; Eva Lee, James Rydge and Su-Lin Garbett-Shiels from Professor Stern's office; Robin Horwitz from the United Nations Foundation; Ailis Rego at the United Nations Convention on Biological Diversity; Karsten Sachs from the German Federal Ministry for the Environment, Nature Conservation and Nuclear Safety; Lisa Pace at the Food and Agriculture Organization (FAO); Erik Haites from Margaree Consultants; Heleen de Coninck and Xander van Tilburg at ECN; Eric Massey and Laurens Bouwer at IVM. We would like to thank UNEP for their support for this book and in particular thanks to John Scanlon and Anne-France White. Finally we would like to thank John Charles the designer of the cover of the book and its companion book, *Human and Environmental Security*.

This book is dedicated to the memory of Joke Waller Hunter (15 November 1946 – 14 October 2005). Joke was a champion for sustainable development. She was the first Director for the Division on Sustainable Development that services the UN Commission on Sustainable Development, a position in which she served from 1994 until 1998. From 1998 to 2002 she was director of the OECD Environment Directorate. She then joined the UN Framework Convention on Climate Change as Executive Secretary. Joke brought people together and through the first years of the UN Commission on Sustainable Development played a critical role in enhancing stakeholder engagement in the UN system. She is missed by many of us.

The book was inspired by listening to Maurice Strong articulate in Tokyo in 2001 the new challenges that will face us in the future in the area of Environmental Security.

> *The important thing is not to stop questioning. Curiosity has its own reason for existing.* (Albert Einstein)

List of Acronyms, Abbreviations and Other Special Terms

ACC	Administrative Committee on Coordination (UN Secretary General's Cabinet)
ADB	Asian Development Bank
AIMS	Africa, Indian Ocean, Mediterranean and South China Sea
AMCEN	African Ministers Conference on Environment
AMU	Arab Maghreb Union
Annex I	the 40 countries plus the European Economic Community listed in Annex I of the UNFCCC that agreed to try to limit their GHG emissions (developed countries)
Annex II	the countries listed in Annex II to the UNFCCC that have a special obligation to provide financial resources and facilitate technology transfer to developing countries (Annex II parties include the 24 original OECD members plus the EU)
Annex B	list in the Kyoto Protocol of 38 countries plus the European Community that agreed to QELRCs (emission targets), along with the QELRCs they accepted (the list is nearly identical to the Annex I parties listed in the convention except that it does not include Belarus or Turkey)
AOSIS	Alliance of Small Island States (42 members and 4 observers: American Samoa, Guam, Netherlands Antilles and the US Virgin Islands)
AWG-LCA	Ad hoc Working Group on Long-term Cooperative Action
AWG-KP	Annex I Parties under the Kyoto Protocol
BASD	Business Alliance for Sustainable Development
BAU	business as usual
BEP	Prince of Wales's Business and Environment Programme
BIPP	bankable investment project profile
BREEAM	Building Research Establishment Environmental Assessment Method
BT	British Telecom
BTL	biomass to liquid
bureau	the bureau of a commission is composed of the *chair* and representatives of the other five *regional groupings* of member states

°C	degrees Celsius
C40	a grouping of key cities around the world to tackle climate change
CAADP	Comprehensive Africa Agriculture Development Programme
$CaCO_3$	calcium carbonate
Cal/EPA	California Environmental Protection Agency
CAN	Climate Action Network
CARICOM	Caribbean Community and Common Market countries (Antigua and Barbuda, Bahamas, Barbados, Belize, Dominica, Grenada, Guyana, Jamaica, Saint Kitts and Nevis, Saint Lucia, Saint Vincent and the Grenadines, Trinidad and Tobago)
CBD	Convention on Biological Diversity
CBI	Confederation of British Industry
CCCI	Cities in Climate Change Initiative
CCD	Convention to Combat Desertification
CCOO	Confederación Sindical de Comisiones Obreras
CCS	carbon capture and storage
CDM	Clean Development Mechanism
CDM EB	Clean Development Mechanism Executive Board
CDP	Carbon Disclosure Project
CDS	chief of defence staff
CEB	United Nations Chief Executives Board
CEE	Central and Eastern Europe
CEIF	Clean Energy Investment Framework
CEO	chief executive officer
CER	certified emission reduction
CFC	chlorofluorocarbon
CGIAR	Consultative Group on International Agriculture Research
CGT	Confederación General del Trabajo
chair	the chair is responsible for facilitating progress in the work of the meeting; in a commission they serve from the end of the previous session until the end of the one that they are chairing (different chairs may be elected for other *informal groups*)
CHP	combined heat and power
CIF	Climate Investment Fund
CLC	Canadian Labour Congress
CLG	Corporate Leaders Group on Climate Change
CNAS	Center for a New American Security
CNG	compressed natural gas
CO_2	carbon dioxide
CO_2e	carbon dioxide equivalent
COMESA	Common Market for Eastern and Southern Africa
COP	Conference of the Parties (to a convention)
CORE	Corporate Responsibility Coalition
CPF	Carbon Partnership Facility
CPSL	University of Cambridge Programme for Sustainability Leadership

CSD	United Nations Commission on Sustainable Development (53 member state governments make up the commission, which meets annually; observer states and non-members, such as the EU, are also permitted to attend)
CSO	civil society organization
CTF	Clean Technology Fund
CTL	coal to liquid
DCDC	Development Concepts and Doctrine Centre
Defra	UK Department for Environment, Food and Rural Affairs
DFID	UK Department for International Development
DRI	Disaster Risk Index
DTI	UK Department of Trade and Industry
DTIE	United Nations Division of Technology, Industry and Economics
EBRD	European Bank for Reconstruction and Development
EC	European Commission
ECA	United Nations Economic Commission for Africa
ECE	United Nations Economic Commission for Europe
ECN	Energy Research Centre of the Netherlands
ECOSOC	United Nations Economic and Social Council
ECOWAS	Economic Community of West African States
EEPS	Energy Efficiency Portfolio Standard
EEZ	exclusive economic zone
EGTT	Expert Group on Technology Transfer (of the UNFCCC)
EIR	Extractive Industries Review
EIT	economy in transition (predominately refers to countries in Central and Eastern Europe)
ENB	*Earth Negotiations Bulletin*
EPA	US Environmental Protection Agency
EPOC	Environmental Policy Committee (of the OECD)
ERA	Environmental Rights Action (Nigeria)
EREC	European Renewable Energy Council
EST	environmentally sound technology
ETF-IW	Environmental Transformation Fund – International Window
ETS	Emissions Trading Scheme
ETUC	European Trade Union Confederation
EU	European Union (works as a group at international meetings, in addition to the actions of the European member states present; presidency rotates every six months)
FAO	United Nations Food and Agriculture Organization
FCPF	Forest Carbon Partnership Facility
FDI	foreign direct investment
FIELD	Foundation for Environmental Law and Development
FoEME	Friends of the Earth Middle East
FPSO	floating production storage and offloading
FRGS	Fellow of the Royal Geographical Society
FRSA	Fellow of the Royal Society of Arts

FSC	Forest Stewardship Council
g	gram
G5	Group of Five (Brazil, China, India, Mexico and South Africa)
G8	Group of Eight developing countries of the Northern Hemisphere consisting of Canada, France, Germany, Italy, Japan, Russia, the UK and the US
G77 and China	the Group of 77 and China was the original group of the so-called non-aligned states; it is, in effect, the negotiating bloc of the member countries present and seeks to harmonize the negotiating positions of its 133 developing country members
GCCA	Global Climate Change Alliance
GDP	gross domestic product
GEEREF	Global Energy Efficiency and Renewable Energy Fund
GEF	Global Environment Facility (the World Bank, UNDP and UNEP established the multi-billion dollar GEF in 1990 to fund environmental programmes, especially in the South and in EITs)
GEO4	*Global Environmental Outlook 4*
GHG	greenhouse gas
GNI	gross national income
GNP	gross national product
GRULAC	Group of Latin America and Caribbean Countries (regional country grouping used within the UN)
Gt	gigatonne
GtC	gigatonnes of carbon
GWh	gigawatt hours
ha	hectare
HDI	Human Development Index
HDR	*Human Development Report*
high-level segment	ministerial-level part of the meeting where the most significant issues are decided
HLCP	High-Level Committee on Programmes (within the CEB)
HoCoN	Host Communities Network
HVAC	heating, ventilating and air conditioning
IAASTD	International Assessment of Agricultural Knowledge, Science and Technology
IAEA	International Atomic Energy Agency
IBRD	International Bank for Reconstruction and Development
ICEM	International Federation of Chemical, Energy, Mine and General Workers' Unions
ICESCR	International Covenant on Economic, Social and Cultural Rights
ICFTU	International Confederation of Free Trade Unions
IDA	International Development Association

IDB	Inter–American Development Bank
IEA	International Energy Agency
IFAD	International Fund for Agricultural Development
IFC	International Finance Corporation
IFI	international financial institution
IGAD	Intergovernmental Authority on Development
IIRSA	Infrastructure Integration for the Region of South America initiative
IISD RS	International Institute for Sustainable Development Reporting Services
ILO	International Labour Organization
IMF	International Metalworkers Federation
IMF	International Monetary Fund
intersessional	official between-sessions meeting of one of the commissions
IOM	International Organization for Migration
IOU	investor-owned utility
IPCC	Intergovernmental Panel on Climate Change
IRR	internal rate of return
ISTAS	Instituto Sindical de Trabajo, Ambiente y Salud
ITUC	International Trade Union Confederation
JI	joint implementation
JISC	Joint Implementation Supervisory Committee
JSE	Johannesburg Stock Exchange
JTF	joint task force
km	kilometre
kW	kilowatts
kWh	kilowatt hours
LCFS	Low Carbon Fuel Standard
LDC	least developed country (49 countries)
LDCF	Least Developed Country Fund
LED	light-emitting diode
LIEE	Low-Income Energy Efficiency programme
LSE	London School of Economics
LTMS	long-term mitigation scenario
m	metre
M&S	Marks & Spencer
major groups	term used in Agenda 21 to describe nine sectors of society fundamental to achieving sustainable development (women; children and youth; indigenous people; NGOs; local authorities; workers and trade unions; business and industry; scientific and technological communities; farmers)
MDB	multilateral development bank
MDG	Millennium Development Goal
MEA	United Nations Millennium Ecosystem Assessment
MEA	Multilateral Environmental Agreement
member state	nation that is a member of the United Nations

mm	millimetre
MIT	Massachusetts Institute of Technology
MoD	Ministry of Defence
MOP	Meeting of the Parties
MOU	memorandum of understanding
MRV	measurable, reportable and verifiable
MW	megawatt
NAFTA	North American Free Trade Agreement
NAMA	nationally appropriate mitigation action
NAPA	National Adaptation Programme of Action
NASA	National Aeronautics and Space Administration
NAST	National Assessment Synthesis Team
NEA	Nuclear Energy Agency
nef	New Economics Foundation
NGO	non-governmental organization
NHS	UK National Health Service
NMTIP	national medium-term investment programme
NORAD	North American Aerospace Defense Command
North	current popular term to describe developed industrialized countries.
NRG4SD	Network of Regions for Sustainable Development
NWF	National Wildlife Federation
NWMO	Nuclear Waste Management Organization
ODA	official/overseas development assistance
ODA	Overseas Development Agency
OECD	Organisation for Economic Co-operation and Development
OES	Bureau of Oceans, International Environment and Scientific Affairs
ONA	Office of Net Assessment
OPEC	Organization of the Petroleum Exporting Countries
OPIC	Overseas Private Investment Corporation
PA	Palestinian Authority
PANPP	Pan-African Non-Petroleum Producers Association
PFAN	Private Financing Advisory Network
ppm	parts per million
PPP	public–private partnership
PR	public relations
PSI	Public Services International
PWA	Palestinian Water Authority
QELRC	Quantified Emission Limitation and Reduction Commitment
QELRO	Quantified Emission Limitation and Reduction Objective
R&D	research and development
REDD	reducing emissions from deforestation in developing countries
regional groups	five regional groups meet privately to discuss issues and nominate *bureau* members and other officials; the regional groups are Africa; Asia; Central and Eastern Europe (CEE);

	Group of Latin America and Caribbean Countries (GRULAC); and the Western Europe and Others Group (WEOG)
RGGI	Regional Greenhouse Gas Initiative
RSB	Roundtable on Sustainable Biofuels
RSPO	Roundtable on Sustainable Palm Oil
RTRS	Roundtable on Responsible Soy
RUSI	Royal United Services Institute for Defense and Security Studies
SADC	Southern African Development Community
SBI	Subsidiary Body for Implementation
SBSTTA	Subsidiary Body on Scientific, Technical and Technological Advice
SCCF	Special Climate Change Fund
SCF	Strategic Climate Fund
SECCI	Sustainable Energy and Climate Change Initiative
SF	Stakeholder Forum for a Sustainable Future
SIDS	small island developing states (especially important in relation to the Barbados Plan of Action for SIDS)
SIPRI	Stockholm International Peace Research Institute
SIWI	Stockholm International Water Institute
South	current popular term to describe developing countries
SRI	socially responsible investment
SUD-Net	Sustainable Urban Development Network
SUV	sports utility vehicle
TJM	Trade Justice Movement
TUAC	Trade Union Advisory Committee (to the OECD)
UEMOA	Union Economique et Monétaire Ouest Africaine
UGT	Unión General de Trabajadores
UK	United Kingdom
UN	United Nations
UNCED	United Nations Conference on Environment and Development
UNCLOS	United Nations Convention on the Law of the Sea
UNCTAD	United Nations Conference on Trade and Development
UNDESA	United Nations Department of Economic and Social Affairs (deals with many of the commissions that review implementation of the outcomes from UN summits and conferences)
UNDP	United Nations Development Programme
UNDSD	United Nations Division for Sustainable Development (division of UNDESA, concerned with sustainable development issues; acts as the secretariat for the CSD)
UNEP	United Nations Environment Programme
UNESCO	United Nations Educational, Scientific and Cultural Organization
UNF	United Nations Foundation
UNFCCC	United Nations Framework Convention on Climate Change
UNGA	United Nations General Assembly
UNGASS	United Nations General Assembly Special Session (usually refers to the 19th Session that was the five-year review of UNCED)

UN-Habitat	United Nations Human Settlements Programme (formerly UNCHS (Habitat))
UNHCR	United Nations High Commissioner for Refugees
UNICEF	United Nations Children's Fund
UNU	United Nations University
US	United States
USAID	US Agency for International Development
USWA	United Steel Workers Association (US)
VC	venture capital
WBCSD	World Business Council for Sustainable Development
WBGU	German Advisory Council on Global Change (Wissenschaftlicher Beirat der Bundesregierung Globale Umweltveränderungen)
WCED	World Commission on Environment and Development
WCI	Western Climate Initiative
WEDO	Water and Environmental Development Organization
WEF	Water Environment Federation
WEOG	Western Europe and Others Group (regional country grouping within the UN; includes the US and Canada)
WFP	World Food Programme
WHO	World Health Organization
WMO	World Meteorological Organization
working group	sub-group of the main meeting, tasked with drafting language for the final documents
WSSCC	Water Supply and Sanitation Collaborative Council
WSSD	World Summit on Sustainable Development
WTO	World Trade Organization
WWF	World Wide Fund for Nature (formerly the World Wildlife Fund)

Part I
Energy Insecurity: Challenges to Future Energy Stability

1
The Challenge of Climate Change from a Developing Country

Marthinus van Schalkwyk

Introduction

We live in interesting and very challenging times. Governments have to really start taking decisions for the long term if we are to survive this century in any meaningful way.

In this chapter, I want to first look at what needs to be addressed internationally – the recommendations at the Hokkaido G8 Summit in 2008 from what has been termed the G5 (Brazil, China, India, Mexico and South Africa) – and then how South Africa is trying to lead by example.

In 1992, the United Nations Climate Change Convention was negotiated as a framework, an umbrella that establishes the key principles and areas of work. In 1995, governments around the world took decisive action and agreed to develop a new legal instrument, the Kyoto Protocol, as a first, but not sufficient, step under this framework.

Adopting another instrument that will coexist with the Kyoto Protocol under this convention framework, parts of which need to be legally binding, may well be what is required next to achieve a balance of interests.

First, we must give legally binding force to mitigation commitments by the one developed country that has not yet ratified the Kyoto Protocol. All eyes are now on the new administration in the US. They must make commitments comparable to the mid-term targets negotiated under the Kyoto Protocol for the second commitment period, and these must be supported by a robust compliance system. The absolute reduction targets for Annex I parties, in aggregate, should be towards the upper end of the range of 25 to 40 per cent below 1990 levels by 2020, and should underpin a long-term goal of domestic emission reductions in developed countries of 80 to 95 per cent below 1990 levels by 2050.

Second, we must create a binding regime for delivery by developed countries of technology, financing and capacity-building support for developing country action, which is measurable, reportable and verifiable. Predictable funding and technology flows to developing countries hold the potential to trigger commensurate nationally appropriate mitigation actions. This will ensure that we may bend the curve of our emissions to deviate substantially from our business-as-usual emission trajectories.

Finally, we must create a mechanism for international recognition of developing country action and to match actions with commensurate incentives.

We must ensure that global emissions peak in the next 10 to 15 years. Otherwise the world will become increasingly insecure. It is now time to raise the bar for all, albeit in a differentiated way for developed and developing countries. South Africa stands ready to do its part.

G5 leadership

As I mentioned earlier, in order to move forward we need the only member of the G8 that has yet to commit to Quantified Emission Limitation and Reduction Objectives (QELROs) to do it soon. From the US we expect comparability of efforts, by which we understand comparability of targets and of compliance, captured in a legally binding manner under the convention. The US must negotiate its commitment together with all nations, and these commitments must be encoded in US domestic legislation. We cannot accept anything that suggests that, because the US has done so little for so long, we must allow them to do less than what is required by science in future.

The signals from President Obama have been encouraging, even though in substance the new administration is still on a *zero* reduction below 1990 levels by 2020. This is clearly not acceptable. Although this is an opening bid, the US would need to come forward with a meaningful negative percentage soon. The same applies to Japan, Russia and Canada, and to the very disappointing announcements on mid-term targets by Australia.

During 2008, the G8 proposed a long-term global goal for emission reductions of 50 per cent by 2050, without a base year and without mid-term targets. We would argue that this proposal does not meet the required-by-science criteria. Without a base year it has no credibility. It is based on the lowest common denominator in the G8 grouping.

During the G8 Summit the G5 have made a much more detailed and more ambitious proposal, also supported by many G8 countries, which essentially has three elements:

1 Developed countries should take the lead with ambitious and absolute emission reductions of between 80 and 95 per cent below 1990 levels by mid century.
2 Quantified emission targets under the Kyoto Protocol should be towards the upper end of the range of 25 to 40 per cent below 1990 levels by 2020 for all developed countries.

3 There should be deviation from business-as-usual emission trajectories in developing countries, supported and enabled by technology and financing.

For South Africa this comes as a political package. To be meaningful, a long-term goal must have a base year, it must be underpinned by clear, unambiguous and ambitious mid-term targets for developed countries, and it should be based on an equitable burden-sharing paradigm that reflects historical responsibility for the problem. Without a base year and mid-term targets, a commitment to '50 per cent by 2050' is an empty slogan without substance.

We were disappointed that some G8 countries continue to reject the ambitious package that we strive for. We would have liked to see much more from the G8. If they accept the G5 package, it would pave the way for significant progress in the current climate negotiations. The developing world gave two steps forward. The ball is now in the G8's court to respond and to deal with the reality of the lowest common denominator between them.

Leading by example

The South African government launched its long-term mitigation scenario (LTMS) process on climate change in 2006. Findings and policy recommendations based on the LTMS were presented to the cabinet in July 2008. This is the culmination of two and a half years of work that involved stakeholders from government, business, civil society and labour.

During the Cabinet Lekgotla, the South African government discussed the policy implications of the LTMS in detail. In response, it has outlined an ambitious vision and adopted a proactive and scientifically and economically robust policy framework that will ensure we meet the challenges of climate change in decades to come. It has set the strategic direction for climate action in South Africa.

Our government's vision and the implementation of this policy framework will be the best insurance policy that current and future generations will have against the potentially devastating impacts of climate change. By adopting this strategic direction, South Africa demonstrates that it is ready to shoulder its fair share of responsibility as part of an effective global response.

The worst impacts of climate change can be avoided if the rest of the world takes up the challenge in a similarly serious way, with developed countries taking the lead.

The international negotiations on strengthening the climate regime after 2012 gained significant momentum at the talks in Bali in December 2007. This process is scheduled to conclude in Copenhagen at the end of 2009.

Science tells us that the climate challenge is urgent and our government has therefore formulated a comprehensive domestic response based on the best available science, scenario-building tools, rigorous analysis of energy and non-energy emissions, the consideration of a wide range of mitigation options and potentials, adaptation planning, and economic models. This is, indeed, cutting edge work.

The world faces a global climate emergency. It is now clear that only action by both developed and developing countries can prevent the climate crisis from deepening. While developed countries bear most of the responsibility for causing the problem to date, developing countries – including South Africa – must face up to our responsibility for the future. While we have different historical responsibilities for emissions, we share a common responsibility for the future.

The technical work done in the LTMS makes it clear that without constraints our emissions might quadruple by 2050. This is, in the most literal sense, not sustainable: if we continue with business as usual, we will go out of business.

The implementation of a combination of the three LTMS strategic options – in other words, those that can be achieved with known technologies and at a relatively affordable cost – can deliver a *substantial deviation from business-as-usual emission trajectories* in South Africa. By committing to and implementing this vision and policy framework, the South African government will make a meaningful contribution to the international effort.

The South African government's vision

The South African government has a vision for its climate policy. It believes that in designing our policy for the transition to a climate-resilient and low-carbon economy and society, we must balance our mitigation and adaptation response. Our climate response policy will be informed by what is *required by science*.

We will continue to proactively build the knowledge base and our capacity to adapt to the inevitable impacts of climate change, most importantly by enhancing early warning and disaster reduction systems and in the roll-out of basic services, water resource management, infrastructure planning, agriculture, biodiversity and in the health sector.

Greenhouse gas (GHG) emissions must peak, plateau and decline. This means that they must stop growing at the latest by 2020 to 2025, stabilize for up to ten years and then decline in absolute terms.

Over the long term, we will redefine our competitive advantage and structurally transform the economy by shifting from an energy-intensive to a climate-friendly path as part of a pro-growth, pro-development and pro-jobs strategy.

Overall, we believe this would constitute a fair and meaningful contribution to global efforts. We would demonstrate leadership in the multilateral system by committing to a substantial deviation from the baseline, enabled by international funding and technology.

The South African government has developed scenarios for mitigation based on key elements. The *Start Now* strategic option as outlined in the long-term mitigation scenarios is based, amongst others, on accelerated energy efficiency and conservation across all sectors, including industry, commerce, transport and residential, inter alia through more stringent building standards.

We will invest in the *Reach for the Goal* strategic option by setting ambitious research and development targets focusing on carbon-friendly technologies, identifying new resources and affecting behavioural change.

Furthermore, regulatory mechanisms as set out in the *Scale Up* strategic option will be combined with economic instruments under the *Use the Market* strategic option, with a view to the following:

- Setting ambitious and mandatory (as distinct from voluntary) targets for energy efficiency and in other sub-national sectors. In the next few months each sector will be required to do work to enable it to decide on actions and targets in relation to this overall framework.
- Based on the electricity-crisis response, government's energy efficiency policies and strategies will be continuously reviewed and amended to reflect more ambitious national targets aligned with the LTMS.
- Increasing the price on carbon through an escalating CO_2 tax or an alternative market mechanism.
- Diversifying the energy mix away from coal while shifting to cleaner coal by, for example, introducing more stringent thermal efficiency and emissions standards for coal-fired power stations.
- Setting targets for electricity generated from both renewable and nuclear energy sources.
- Laying the basis for a net zero-carbon electricity sector in the long term.
- Encouraging renewable energy through feed-in tariffs.
- Exploring and developing carbon capture and storage (CCS) for coal-fired power stations and all coal-to-liquid (CTL) plants, and not approving new coal-fired power stations without carbon capture readiness.
- Introducing industrial policy that favours sectors using less energy per unit of economic output and building domestic industries in these emerging sectors.
- Setting ambitious and, where appropriate, mandatory national targets for the reduction of transport emissions, including through stringent and escalating fuel efficiency standards, facilitating passenger modal shifts towards public transport, and the aggressive promotion of hybrids and electric vehicles.

Process going forward: 2009 to 2012

The South African government has mandated a clear path for the future. Milestones include a national summit held in March 2009, the conclusion of international negotiations at the end of 2009, and a final domestic policy to be adopted by the end of 2010 after international negotiations have been completed.

The process will culminate in the introduction of a legislative, regulatory and fiscal package to give effect to the strategic direction and policy from now until 2012.

South African business response

In November 2008 I spoke at the launch of the second round of the South African contribution to the Carbon Disclosure Project (CDP). We found that the businesses that contributed to the project have recognized that climate change and its impacts are the biggest environmental challenges we are facing in this century, and they have decided to be part of a collective response. They understand that it would not be economically, environmentally or politically sustainable for South Africa to continue to increase our emissions along a business-as-usual path.

To tackle the challenges ahead, definitive leadership within both the public and private sectors will be required, particularly given competing social and economic priorities, comparatively high per capita GHG emissions and high vulnerability to the impacts of climate change.

This second CDP survey found encouraging evidence that South African companies are beginning to appreciate and respond meaningfully to this challenge:

- This year, the CDP was extended to include the top 100 companies on the Johannesburg Stock Exchange (JSE). This means that the sample size more than doubled since last year's report, which only included the JSE top 40.
- South Africa's second CDP generated a response rate of 59 per cent.
- There has been a sizeable increase in the number of companies disclosing their GHG emissions. More than 75 per cent of responding companies disclosed their GHG emissions this year. While in several instances this disclosure is only on a partial basis, there is, nevertheless, an emerging commitment to improved monitoring and reporting on greenhouse gas emissions.
- In a significant change on last year's responses, there appears to be much greater awareness of, and engagement in, government policy on climate issues. This builds on the long-term mitigation scenarios process, which entailed the engagement of numerous senior executives across different sectors.

A collaborative approach to climate solutions is developing. Acknowledging that an escalating price on carbon will become part of the future business environment, several companies are beginning to work with government, stakeholders and industry peers on national policies relating to climate change. Notwithstanding these encouraging developments, there remains room for improvement:

- Relatively few companies (23 per cent) have disclosed specific company-wide GHG emissions reduction targets, and most of those companies that have emissions targets have focused on reducing their emissions intensity, rather than striving for a reduction in absolute emissions. If South Africa's emissions are to peak and then decline, companies will need to demonstrate a significantly higher level of ambition.

- On a related issue, while most responding companies have developed, or are implementing, formal systems for measuring and reporting on their GHG emissions, some important gaps remain in their governance systems for climate change, and in the nature and extent of executive board oversight on this issue.
- Carbon-intensive companies dominate South Africa's disclosed GHG emissions. The three largest GHG emitters (excluding ESKOM) – Sasol, BHP Billiton and Anglo American – account for two-thirds of the total reported GHG emissions of responding listed companies in South Africa.
- Electricity consumption constitutes 41 per cent of the total reported GHG emissions.

It is evident that South Africa's current electricity generation mix has a profound impact upon the majority of corporate emission levels. Although there have been some encouraging improvements since last year, there are still indications that mainstream South African investors do not fully appreciate the business implications of climate change, or that they are exerting meaningful influence on the corporate sector on this issue.

South Africa has remained sheltered from emission reduction targets under the Kyoto Protocol due to its status as a developing economy. As a result of this, our private sector has been slower than some of its counterparts in learning how it will be affected by climate change.

We need to be prepared as the future climate regime will see South Africa having to make a substantial deviation of its emissions trajectory below baseline – in a measurable, reportable and verifiable way.

It is hoped that climate change awareness through the CDP will facilitate appropriate decisions amongst those companies involved. New opportunities and incentives could emerge for those who are proactively positioning themselves to face the changing political and economic landscape in a carbon-constrained world.

Not only does proper tracking and reporting make business sense, but it is only when companies know their carbon footprint that they can properly plan to mitigate. It is also an indicator of good corporate governance, of accountability and of taking co-ownership for the future.

Government is determined to continue to take responsible action on climate change. I am therefore looking forward to working even more closely with business in building on the work of the LTMS and the CDP. It is only by working in partnership that we will be able to make the kind of progress that is required by the latest scientific findings on climate change. Responding early and decisively is the biggest contribution that business can make.

The National Business Initiative and Incite Sustainability, the authors and sponsors of the report and companies see that the CDP can help in unlocking new opportunities, creating new competitive advantages in new products, technologies and processes, and instigating a shift to greater efficiency in existing operations. It is vital that those companies that do not yet track, report and manage their emissions join these efforts. It will place them in a position to understand their future risks and vulnerabilities in a changing global and domestic business environment and to be prepared for an era in which mandatory reporting of GHG emissions will become part of the landscape.

Conclusions

The challenges we will face in the coming years because of climate change and its impacts upon our insecurity will be great. We know that the industrialized model developed countries have used over the last 150 years isn't sustainable. The changes we need to make in society are immense.

Some people talk about environmental technology being the next wave after the information age. It's clear that we will need new technology and we will have to test many new ideas; but perhaps the largest changes will have to be in ourselves. We need to recognize that only by working together, people, communities and nations can stand any chance of creating a sustainable planet for all.

We will need great leaders who can help us through these difficult times. I am reminded of the words of former South African President Nelson Mandela in his biography *Long Walk to Freedom*:

> *I have walked that long road to freedom. I have tried not to falter; I have made missteps along the way. But I have discovered the secret that after climbing a great hill, one only finds that there are many more hills to climb. I have taken a moment here to rest, to steal a view of the glorious vista that surrounds me, to look back on the distance I have come. But I can rest only for a moment, for with freedom comes responsibilities, and I dare not linger, for my long walk is not yet ended.*

2
Energy Consumption Challenge of the Industrialized World: An Overview of German Climate Policy

Sigmar Gabriel

Climate change is potentially more dangerous than all the conflicts put together. (Ban Ki-moon, UN Secretary General, 1 March 2007)

Climate change is here

Climate change is a reality. Observations from all continents of the Earth are providing an increasingly precise picture of dramatic change. Global temperatures have increased by 0.8°C since the beginning of industrialization. Sea levels rose twice as much in the period from 1993 to 2003 as in the three previous decades. Icebergs and glaciers are melting at a record pace; permafrosts that have been stable for centuries are thawing. Extreme weather events such as floods, severe hurricanes, heavy rainfalls and major heat waves have become much more frequent. Climate change is not only a threat to the natural basis of life; it is also a threat to global peace and security. If the glaciers in the Himalayas melt, 40 per cent of Asia's drinking water supply is at risk. As a consequence, there will be increased movements of refugees and armed conflicts over water. Already today, there are more refugees because of global warming than because of war and civil war.

The science is clear – the anthropogenic greenhouse effect is the main cause of climate change. The Intergovernmental Panel on Climate Change (1990, 1995, 2001, 2007) has proven this with an abundance of scientific data and warns us that the speed of global warming has increased – it is advancing even more rapidly than previously assumed. Mankind has only 10 to 15 years to prevent the worst impacts and to respond adequately to this huge challenge. If climate change remains unchecked, the global temperature is predicted to increase by 6°C by 2100 – with disastrous consequences.

Decision-makers are therefore called upon to take serious action immediately. We have a responsibility to the people in our countries and worldwide, as well as to present and future generations to preserve the climate conditions to which life on this planet and human civilization have adapted over millennia. Our duty to act is all the more compelling as the technology to avert climate change – and this is the good news – is already available. The challenge, therefore, is quite simply to create the necessary conditions for the widespread deployment of this technology. Europe and Germany's goal is to limit the global temperature rise to 2°C compared to pre-industrial times, a level at which the consequences of climate change remain just about manageable according to climatologists.

Climate protection is affordable, inaction is not

Lord Nicholas Stern, the former chief economist of the World Bank, has shown that, ultimately, neglecting climate protection would lead to devastating consequences (Stern et al, 2006). He placed the global costs of an unchecked climate change at an annual 5 to 20 per cent of the gross world product – an economic damage greater than the impacts of World War I and II and the Great Depression combined. The costs of climate protection, on the other hand, would be only 1 per cent of the gross world product. The question, therefore, is no longer whether we can afford climate policy. The question is: can we afford climate change?

The international response to climate change

However, climate protection can only be achieved if all countries make their contribution. At the Earth Summit in Rio de Janeiro in 1992 (UNFCCC, 1992), the necessity of a global response was first acknowledged internationally. The summit adopted the Framework Convention on Climate Change. In 1997, in the Kyoto Protocol (UNFCCC, 1997) the industrialized countries acknowledged their particular responsibility and accepted binding targets for the reduction of their greenhouse gas emissions by 2012.

The Climate Change Conference in Bali in December 2007 (UNFCCC, 2007) was a decisive milestone for continuing the Kyoto process with ambitious targets beyond 2012. At the G8 Summit in Heiligendamm in June 2007, the major industrialized countries agreed to continue negotiations under the roof of the United Nations. Following the principle of 'common but differentiated responsibility', the G8 countries want to take on a leading role in the reduction of greenhouse gases in this process and committed themselves to halving global greenhouse gas emissions by 2050 as a long-term goal.

Europe: A leading power in climate protection

Under the German European Union (EU) presidency (German Government, 2007), in March 2007 the EU heads of state and government adopted a far-reaching and, in fact, historic decision on future climate policy: by 2020, EU greenhouse gas emissions will be reduced by 20 per cent below their 1990 level, and even by 30 per cent if other industrialized countries adopt similarly ambitious targets and newly industrializing countries contribute adequately.

To achieve this target the EU has decided that by 2020, 20 per cent of Europe's energy consumption will be covered by climate-friendly renewable energies, with biofuels having at least a 10 per cent share in petrol and diesel. At the same time, energy efficiency will be increased in order to achieve the objective of saving 20 per cent of the EU's energy consumption compared to projections for 2020. As a long-term objective, the European Council has specified a 60 to 80 per cent reduction of greenhouse gases in Europe by 2050 compared to 1990.

German climate protection philosophy

Both in the international process and in Europe, Germany will remain a front-runner in climate change policies. Under the Kyoto Protocol (UNFCCC, 1997), Germany has agreed to reduce its greenhouse gas emissions by 21 per cent by 2012 compared to 1990, and is now well on the way to accomplishing this goal. Accordingly, for 2020, Germany is aiming at ambitious emissions reductions: Germany will reduce our greenhouse gases by 30 per cent, or even by 40 per cent in the case of similar ambitious commitments by the other industrialized countries and the newly industrializing countries.

However, the implementation of such goals requires a fundamental modernization of our industrial society. It requires the transition from the old fossil-based energy industry with high energy consumption levels to a modern low-carbon society based on renewable energies and high energy efficiency – in other words, nothing less than a third industrial revolution.

To meet this end, economic growth must be decoupled from the emission of greenhouse gases. Furthermore, current emission levels must be reduced even further. As a crucial framework condition, we need an international agreement on binding targets to ensure a fair distribution of the costs of climate protection and to avoid lopsided disadvantages in worldwide competition. At national levels, for the implementation of these targets we need instruments and measures which accelerate technological progress and the deployment of the available sustainable technologies.

Here, market forces play a decisive role. However, in the past we have seen a major malfunctioning of markets. Due to energy prices that either did not reflect the true costs of greenhouse gas emissions or were not fully transparent to consumers, clean technologies were outrivaled by the old, cheaper but energy inefficient technologies. Therefore we cannot just rely on research funding to trigger the necessary innovations. However, climate change policies that work against the markets will not succeed. Governments therefore need to strive to 'get the prices right' and in this way unleash market forces as the most powerful ally in following a new climate friendly path. With a fixed emissions cap for major emitters and the establishment of an emissions trading scheme, the EU has made an important step in this direction, laying the basis for an international carbon market.

But more can and must be done to activate market forces. Success will only be achieved through a combination of additional economic incentives for industry and consumers, consumer information, and the promotion of new technology. Where markets still fail, regulatory intervention, particularly minimum standards and monitoring by public agencies, becomes inevitable.

Reaching a fair international agreement with binding targets for each country, formulating the necessary well balanced policy package that enables our societies to grow and prosper and thus transforming our industrial societies into a low-carbon society – that is the challenge for the industrialized world, and one which I am sure can be accomplished.

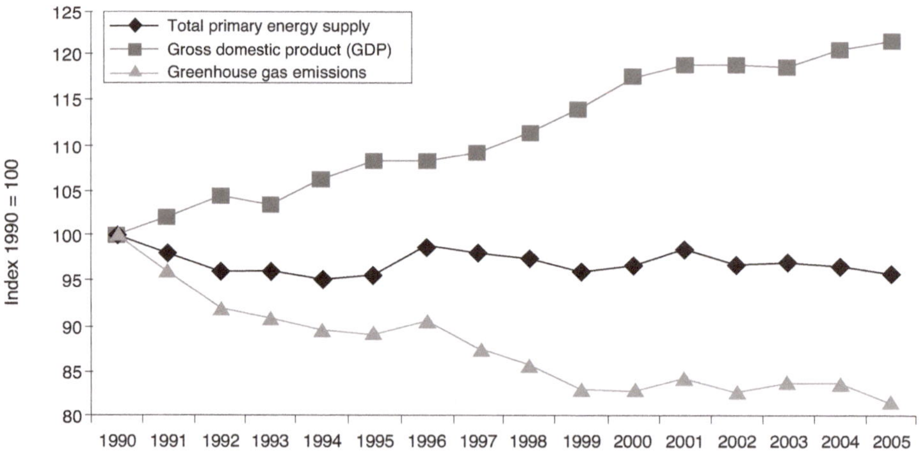

Source: BMU/Destatis (2007)

Figure 2.1 *Decoupling growth, energy consumption and emissions in Germany*

Germany on the road to a low-carbon economy

Germany is making good progress in climate protection, with greenhouse gas reduction currently at 18 per cent, largely meeting the Kyoto target for 2008–2012. The privatization of East German energy supply and its consequent modernization in the 1990s, the Ecological Tax Reform in 1999 (German Government, 2002a), the Renewable Energy Sources Act in 2000 (German Government, 2000), the Combined Heat and Power Act in 2002 (German Government, 2002b) and the launching of emissions trading in 2005 have been important milestones in German climate policy.

Now, reflecting the urgent need for action, the German government has adopted a comprehensive 29-point future-oriented package of measures on energy and climate policy. Federal funding for climate protection is being increased drastically to 2.6 billion Euros, up by 1.8 billion Euros compared to 2005. Never before has Germany had such an extensive programme for climate and environmental protection. The new package of measures will tackle carbon dioxide in all relevant areas and, thus, integrate climate protection into all sectors of society. The package contains 21 acts and ordinances and 10 support programmes and is calculated to bring down German greenhouse gas emissions by more than 30 per cent compared to 1990.

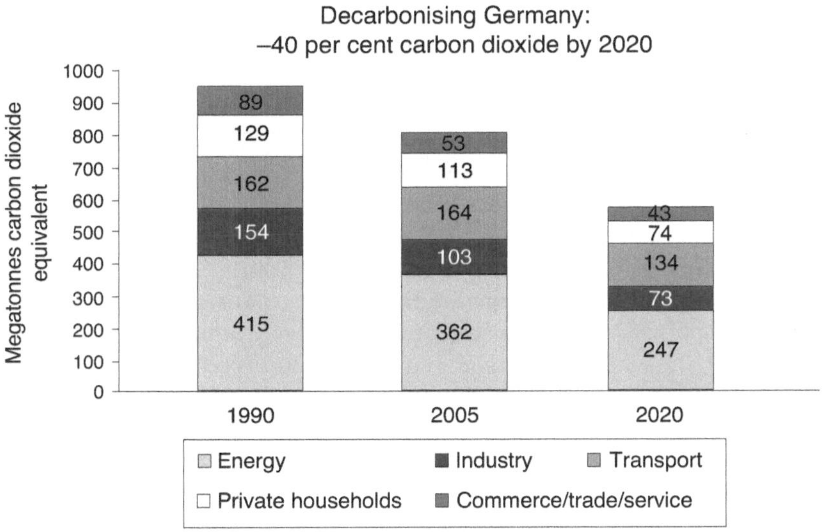

Source: German Federal Environment Agency, 2007

Figure 2.2 *Sector-wide reduction in carbon*

Revolutionizing energy generation

Germany will dramatically restructure its energy supply by 2020. Electricity from solar, wind, biomass and other renewable energies will have a share of 25 to 30 per cent, up from 6.3 per cent in 2000 and around 12 per cent in 2006. The renewables' share in heat supply will rise to 14 per cent, more than twice the current share of 6.3 per cent.

The key to the success so far has been the Renewable Energy Sources Act (German Government, 2004), which guarantees the feed-in of electricity from renewables at a fair and fixed fee. Thus, the additional costs of renewable energies are apportioned among all electricity consumers. The act has since served as a model for more than 30 countries. The expansion of renewables also shows how countries can profit from fostering climate-friendly technologies. This has enabled German industry to win an important competitive margin in a lead market, with positive effects for other sectors too. In terms of employment, more than 400,000 jobs are expected to be created in the field by 2020.

A crucial role on the supply side is also attributed to combined heat and power (CHP) generation. CHP plants use the heat produced for the generation of electricity for the supply of surrounding residential areas or for process heat in industry and thus achieve an outstanding level of efficiency of up to 90 per cent. It is Germany's aim to increase the share of CHP in electricity generation to 25 per cent by 2020 – twice the current level. Combined heat and power plants are supported with fixed fees per kilowatt hour of electricity, similar to the successful feed-in tariffs for renewable energies. The construction of heat pipes will also be supported by investment grants of 20 per cent.

In addition, under the European emissions trading scheme, the German government has considerably lowered the emission cap for the energy-producing sector and energy-intensive industry. In 2008 to 2012, companies will receive allowances about 15 per cent below their present emissions level. Moreover, while in 2005 to 2007 allowances were allocated free of charge, now 10 per cent will be auctioned. The mode of allocation will create an incentive for the energy-producing sector to reduce emissions. It should be also noted that the forthcoming modernization of the German power plant park will see only the most efficient plants being connected to the grid. Revenues ensured from the auctions will be used to promote climate protection research and investments both in Germany and in developing countries.

Finally, the German government supports the development of carbon capture and storage (CCS). With the cooperation of German industry, two or three CCS demonstration power plants will be set up in Germany by 2015. Our aim is that from 2020 at the latest, the construction of CCS power plants will be a standard throughout the world. Only under such conditions will coal-fired power plants still have a future in a low-carbon world.

Nuclear power is intentionally excluded from the German government's climate protection concept. The dangers of nuclear power are well known, and the disposal of highly toxic nuclear waste is still a global problem. Moreover, uranium reserves are expected to run out within the next 35 years. The German government is therefore convinced that nuclear power will only play a small role, if any, in the world's future energy mix. The solution to the pressing questions of the future lies in the massive expansion of renewable energies and a major increase in energy efficiency.

Energy efficiency: Uncovering hidden treasures

Energy efficiency is our best source of energy. Every kilowatt hour of electricity, every litre of oil, every tonne of coal that we do not consume protects the climate and saves money. The implementation of the EU target of a 20 per cent efficiency increase by 2020 alone is estimated to reduce energy costs in Europe by 100 billion Euros per year. Moreover, investments in energy efficiency pay off in many cases with the technology available today. However, awareness of this fact still has to be conveyed to decision-makers in all sectors, to private consumers, managers of small- and medium-sized enterprises and officials responsible for public spending.

Without its own legal competency in the common European market, Germany therefore supports proposals by the European Commission for ambitious minimum efficiency standards, effectively excluding products with higher electricity consumption levels from the market. The European Commission has identified 19 priority product groups to be targeted up to 2009, from fridges, washing machines and tumble dryers to personal computers, televisions, lights and electric motors. Here, Germany calls for a top runner approach: today's best appliances will determine the minimum standard of tomorrow.

At the same time, Germany is urging the EU Commission to introduce a more effective and consumer-friendly labelling of energy-using products. Often consumers are unaware of 'hidden costs' caused by energy consumption in standby mode or the overall life-cycle costs of an appliance, although these are often higher than the actual purchase price. Consequently, consumers need to be better informed to enable them to take energy savings into account in their purchase decisions.

There is also huge potential for energy savings in buildings. Old buildings consume 20 to 30 litres of oil per square metre per year, an amount that could be reduced by more than three-quarters through better insulation and more efficient heating systems. House owners will now be required to exhaust this potential in cases of a basic modernization of the building or heating system. From 2009 on, the design of new buildings must guarantee consumption at 3 litres of oil per square metre. The goal for 2020 is to become completely independent of oil and gas by means of intelligent building technology and renewable energies.

As an incentive to owners to modernize their houses, the German government has quadrupled its funding for modernizing energy systems in buildings compared to 2005, up to a total of US$1.8 billion. There is also increased support for tenants' rights: if the landlord of a property does not modernize it and, as a result, the heating costs are too high, tenants should be allowed to reduce their rent.

Transport: Full speed ahead while saving fuels

The passenger car is the hub of the transport sector. According to the IPCC (IPCC, 2007), by the middle of this century the number of cars worldwide could more than double. It will thus be all the more important to produce greener cars. In view of an oil price that has almost quadrupled since 2001, this will also lead to massive savings for the consumer. Germany therefore supports the plan of the European Union to apply standard carbon dioxide limits to all new cars in Europe by 2012 aiming at a reduction from an average of 170g of carbon dioxide per kilometre today to 120g of carbon dioxide per kilometre. This corresponds to an average consumption of 5 litres of petrol per 100km. Further reductions will be necessary and are planned up to 2020 with a view to the technological progress.

The German government will support this development with economic incentives: in the future, the annual motor vehicle tax will be calculated on the basis of carbon dioxide emissions rather than the size of the vehicle. Similarly, the motorway toll charged for lorries will depend upon the carbon dioxide emissions of the vehicle.

As well as the vehicles themselves, it is also important that fuels become more environmentally friendly. By 2020, refineries will be required to blend an annually increasing share of ethanol and biodiesel with petrol and diesel, up to a final share of 20 per cent. At the same time, sustainability standards will ensure that the cultivation of raw materials is sustainable and does not entail the destruction of valuable forest areas.

At the European level, Germany advocates the incorporation of international air and shipping traffic into emissions trading. The European Union is currently devising corresponding regulations, which will apply from 2011.

The transport concept is rounded off by the promotion of public transport. Germany has a fully developed bus and rail network that receives public funds since bus and rail travel secure climate-friendly mobility.

Climate and energy policy for a secure future

Six and a half billion people live on this planet, and this figure is expected to rise to 9 billion in a matter of years. For these people, growth and prosperity is linked to access to energy. The world, therefore, is facing a twofold and growing challenge: to tackle climate change effectively and to secure energy supply.

Germany, as well as the EU as a whole, is determined to make its contribution. We will implement the climate package before an international agreement is concluded and thus build trust, particularly among the developing countries, that the industrialized countries are recognizing their particular responsibility and that words are being followed by deeds.

We cannot afford a situation in which everyone waits for the others to make the first move. We need to change our attitude: from 'you first' to 'me too'.

References

BMU/Destasis (2007) *Fundamental Aspects of Environment and Energy, Climate Protection*, Federal Environment Ministry, Germany

German Federal Environment Agency (2007) *Decarbonising Germany*, German Federal Environment Agency, Berlin, www.umweltbundesamt.de/index-e.htm

German Government (2000) *The Renewable Energy Sources Act in 2000*, German Government, Berlin, *www.bmu.de/files/pdfs/allgemein/application/pdf/res-act.pdf*

German Government (2002a) *The Effects of Ecological Tax Reform in German: Energy Policy*, German Government, Berlin, *www.bmu.de/files/erneuerbare_energien/downloads/application/pdf/diw_energypolicy2002.pdf*

German Government (2002b) *Combined Heat and Power Act in 2002*, German Government, Berlin, *www.bmu.de/english/international_environmental_policy/johannesburg_summit_2002_/doc/3403.ph*

German Government (2004) *Renewable Energy Sources Act*, German Government, Berlin, *www.bmu.de/english/renewable_energy/6465*

German Government (2007) *German Presidency of the European Union Final Assessment,* German Government, Berlin, www.eu2007.de/en/News/Press_Releases/June/0626BMVBSBilanz.html

IPCC (Intergovernmental Panel on Climate Change) (1990) *IPCC First Assessment Report*, UN, Geneva

IPCC (Intergovernmental Panel on Climate Change) (1995) *IPCC Second Assessment Report*, UN, Geneva

IPCC (Intergovernmental Panel on Climate Change) (2001) *IPCC Third Assessment Report*, UN, Geneva

IPCC (Intergovernmental Panel on Climate Change) (2007) *IPCC Fourth Assessment Report*, UN, Geneva

Ki-moon, B. (2007) Secretary-General's address to UNIS-UN Conference on Climate Change, www.ipb.org/environment.html

Stern, N., S. Peters, V. Bakhshi, A. Bowen, C. Cameron, S. Catovsky, D. Crane, S. Cruickshank, S. Dietz, N. Edmonson, S.-L. Garbett, L. Hamid, G. Hoffman, D. Ingram, B. Jones, N. Patmore, H. Radcliffe, R. Sathiyarajah, M. Stock, C. Taylor, T. Vernon, H. Wanjie, and D. Zenghelis (2006) *Stern Review: The Economics of Climate Change*, HM Treasury, London

UNFCCC (United Nations Framework Convention on Climate Change) (1992) *UN Framework Convention on Climate Change*, UN, New York, NY

UNFCCC (1997) *Kyoto Protocol*, UN, Bonn

UNFCCC (2007) *COP13 Bali Action Plan*, UN, Bonn

3
Oil: How Can Europe Kick the Habit of Dependence?[1]

Derek Osborn

Introduction

Two critical factors will shape the future of oil production and consumption over the next decades. The first is accelerating climate change. The second is the increasing difficulty in finding secure sources of oil in the world. The interaction of these two factors is currently leading the world into a more and more unstable position.

The only successful way out of the climate change crisis will be for the global economy to manage a rapid transformation of its energy base away from its current excessive dependence upon the burning of fossil fuels. The Intergovernmental Panel on Climate Change (IPCC, 2007) has estimated that in order to contain global temperature rises to no more than 2°C to 2.4°C above pre-industrial levels, CO_2 emissions will need to be reduced by 50 to 85 per cent from 2000 levels by 2050. The IPCC and the International Energy Authority (IEA, 2008) have analysed a number of different scenarios for achieving such reductions, some of which differ in detail. But it is clear that in all plausible scenarios each type of fossil fuel will have to play its part in achieving the reductions needed. For oil, in particular, global consumption will need to stop growing within a few years and then to decline steadily to much lower levels by the mid century.

The supply-side situation reinforces this argument. The world's resources of oil are finite and cannot last forever. New sources of oil supply in the world are still being discovered; but the new finds tend to be smaller and more difficult to develop, and are sometimes in politically unstable parts of the world. They may be more expensive to develop, and there will be more competition in the world to secure access to the supplies that are still available.

Some of the new sources are also located in environmentally sensitive areas such as the Arctic. Some new sources, such as the Canadian tar sands, will also be more difficult to operate, and the extraction process will itself produce larger CO_2 emissions. It would be better to avoid using such sources at all if possible, or at least to postpone using them until better environmental and carbon capture safeguards can be put in place.

Taken together, these two factors make an overwhelming case for taking action as soon as possible to check the growth in global demand for oil and to start managing it downwards over the next 50 years. The IPCC and IEA illustrate the kind of radical transformation of current patterns of consumption and production throughout the world that this will require.

Most of the current scenarios do not establish separate pathways for reductions in consumption of the different fossil fuels. In future iterations it might be helpful to establish a generally agreed trajectory or benchmark for the progressive reduction of global demand for oil (and other fossil fuels) over the next 50 years. Since many developing countries have a legitimate expectation that their own consumption of fossil fuels must be expected to continue to rise for some time as their development progresses, it is clear that the developed countries will have to shoulder the main burden of the reductions to be achieved – at least initially. The sooner such a downward trajectory can be firmly established in all government and business strategic planning, the more chance there is that the world may manage a soft landing to the inevitable climate change and oil price shocks ahead.

Europe only accounts for less than 20 per cent of global demand for oil and cannot achieve the necessary global change by its own actions alone. Europe is, however, one of the regions heavily dependent upon imports for its supplies of oil (over 80 per cent) and is therefore among the regions most vulnerable to emerging supply constraints. It does, therefore, have a strong incentive to take a lead in managing the transition to a new low-carbon economy that is less dependent upon oil and other fossil fuels. If it could do so successfully, it could provide a useful example for others.

What needs to be done: Diversifying away from oil

Assuming that oil will have to take a significant share of the burden of reduction, it seems likely that European oil consumption will need to be reduced by well over 50 per cent by 2050. This chapter briefly reviews the kind of changes that will be needed to achieve this reduction, and the kind of measures needed to drive or encourage these changes.

Oil is principally used in the transport sector, but with significant proportions in domestic heating and cooking, in heating and cooling other buildings, in power generation and as a feedstock for the petrochemical industry. In all of these sectors it will be necessary to reduce or eliminate reliance on oil as quickly as possible.

The transport sector

In the transport sector three main changes are needed:

1. Integrate transport and land-use policies for cities, towns and other settlements in order to reduce journey lengths and times wherever possible.
2. Continuously improve the energy efficiency and carbon performance of planes, ships, trains and road vehicles of all kinds, with the introduction of electric- or hydrogen-powered vehicles as soon as possible.
3. Use more carbon-efficient modes of transport for journeys: rail trips rather than air flights; public transport rather than private transport; and cycling and walking wherever possible.

Aviation may have to remain a privileged user of oil, at least for the next two or three decades for essential uses. However, it needs to make the maximum possible efficiency improvements, and expansion of high-speed rail services should be preferred to air transport wherever possible. Further expansion of air transport and airports should also be discouraged.

In relation to shipping, efficiency improvements should be sought continually, and innovative ideas such as adding supplementary wind power to reduce fuel consumption need to be actively encouraged.

Household consumption of oil

Direct burning of fossil fuels in fires, boilers or for cooking will need to be phased out, and electricity (increasingly drawing on locally based renewable generation as well as the grid) or sustainably grown wood will have to become the standard household fuels. It would be useful to establish a timetable for this transition.

Power generation

In terms of power generation there will need to be a great effort to expand renewables of all kinds as rapidly as possible. In Europe the targets that have been set for the expansion of renewables are a good beginning; but more needs to be done to carry the different technologies into the marketplace at affordable prices.

Coal (and, to a lesser extent, other fossil fuels) will remain an important fuel for power generation for several decades ahead. Carbon capture and storage should be developed as soon as possible. It should then be mandated for any remaining oil-fuelled power stations as well.

A new generation of nuclear power plants may also have some part to play. But nuclear power technology has its own sustainability problems, and must not be allowed to divert investment resources and political attention from the major expansion of renewables and energy efficiency that are the primary goal of the transition.

Oil in the business sector

A similar transition will be needed in the industrial and business sector of the economy for all general heating and other purposes. Where industrial processes currently rely on the use of fossil fuels as a feedstock, sector-by-sector analysis will be needed to identify how far carbon emissions from such processes can be captured and stored or where such uses can be substituted by non-fossil fuel processes.

In order to assist all parties concerned to plan for these changes, it would be useful to establish indicative pathways for the levels of savings to be achieved in each sub-sector that uses oil, and the likely timetables for these transitions both at global and regional levels.

What needs to be done: Policy measures to reduce reliance on fossil fuels and to encourage diversification of energy supply

At present, the trajectory of global consumption of oil is still rising from year to year, now principally driven by the rapid expansion of demand in the emerging economies. Although demand in some developed countries is now nearer to stabilizing, no country has yet established the steadily declining trajectory that will be needed.

The market cannot be expected to deliver the changes that are needed without active intervention by governments and a sustained effort to engage the public and organizations of all kinds. Many of the measures necessary to encourage and promote the transformations are already well known, in principle. But everywhere in the world, the range of measures needs to be extended and to be applied more vigorously and urgently.

Three types of measures are needed:

1 fiscal and budgetary measures to encourage the development of low-carbon goods and services, and to discourage high carbon consumption;
2 regulatory measures to the same end;
3 promotional measures to encourage active engagement by the public and organizations of all kinds in bringing about the changes needed.

Fiscal measures to put a proper price on carbon emissions

Fossil fuels should bear the full cost of the burden that the emission of carbon dioxide is imposing on the world. This requires either taxation of carbon products (such as gasoline), or a system for rationing and trading permits to release carbon dioxide, or both.

The European Union (EU) already has comparatively high levels of taxation of oil and oil products, but may still need to go further in this direction. It has also pioneered a system for rationing and trading permits to release carbon; but this system still contains a number of anomalies and loopholes and has not yet established a clear and stable market signal in favour of reducing fossil-fuel consumption and diversifying to other fuels. Anomalies need to be fixed, and exemptions reduced. Above all the system needs to be extended to the rest of the developed world, and as soon as practicable to the emerging economies as well. This should be a key objective in the context of the international climate change negotiations.

Some commentators have suggested that carbon trading systems might be developed in such a way that a floor price for carbon is established to give greater certainty to the market. The floor price might then be driven steadily upwards over the next three decades in order to provide a steadily increasing market pressure on operators of all kinds to diversify away from fossil fuels.

Some of the new technologies necessary are still at the development stage and will need significant support and encouragement from the public sector to secure their early introduction and wide deployment. Carbon capture and storage, further development of renewables, third- and forth-generation renewables, electric- or hydrogen-powered vehicles and the infrastructure they will need are all in this category and require major public support to get them deployed throughout the world as soon as possible.

Major investment in the rail sector is needed to spread electrification and make rail travel the preferred alternative to air flight for most short-haul routes in Europe and in other parts of the world.

Regulatory measures

Fiscal measures are not sufficient by themselves to drive the necessary transition away from oil. Demand is too inelastic, and there are political constraints on driving up the prices of oil products too rapidly. There needs to be a comprehensive programme of regulatory measures to drive up standards and drive out inefficient processes and products. There also needs to be support for the research, development and introduction of new technologies.

In particular, there should be a comprehensive and urgent programme to drive up the energy-efficiency standards of all energy-consuming products and services. Europe has made some progress on this, but still has further to go, both in setting the standards and in ensuring that they are complied with. The heating and cooling of buildings, for example, is still massively inefficient and requires a vigorous programme of action to secure rapid improvement.

Standards for improving the carbon emissions from vehicles are particularly important. New tighter standards have recently been agreed in Europe; but even more stringent standards for future years will soon be needed in order to establish a firm planning framework for the motor industry to adjust to. The next step now urgently called for is to establish a similar tight programme for the progressive improvement of emissions from vans and from heavier goods vehicles. Again action within Europe needs to keep pace with and be matched by comparable efforts in other parts of the world.

Involving civil society

Much more must be done to bring the general public, business, trade unions and other civil society organizations into partnership and participation in the common effort.

Citizens need to be encouraged and incentivized to play their part through such means as improving the efficiency of their homes and cars, using greener forms of energy for lighting and heating, purchasing more energy-efficient goods and services, and reducing the carbon impact of their regular travel and holidays. In many countries there is already a growing proportion of the public and of civil society organizations (CSOs) who would be ready and willing to take action if only they could be given a strong and effective political lead as to what is expected of them, along with appropriate incentives for action.

Many local and regional government bodies have already shown vision and courageous political leadership on this issue. They need to be encouraged and incentivized to go further. Businesses similarly need to be incentivized to make further progress. They should be urged to continuously improve the energy efficiency of their operations and to obtain their energy from low-carbon sources. Regulation should be used more systematically and vigorously to drive up the energy performance of all types of products and services. Industry must devote more effort to help in the transition to a world of lower demand for oil in the future, and less to developing more marginal sources of oil, particularly where these are going to cause severe environmental damage.

The oil industry of the world has a particular responsibility for assisting the transition to a lower carbon economy. It needs to:

- recognize the necessity of the transition of the global economy to a declining usage of fossil oil over the decades ahead, and think, plan and behave accordingly;
- continue to improve the oil industry's own carbon performance in its operations;
- substitute biomass or other carbon-neutral resources for the use of fossil fuels in its products where feasible and sustainable;
- use the oil industry's massive resources of skills and financial power to help other aspects of the transition, and the earliest possible development and deployment of carbon capture and storage; and
- work closely with the automobile industry to help accelerate the transition to low- or zero-carbon vehicles.

Governments should initiate an intensive dialogue with the global oil industry to try to develop a common understanding of the trajectory for declining oil usage that needs to be established, and to provide appropriate incentives to encourage (or, if necessary, compel) the industry to move in these key directions.

There is some scope for biofuels development (and particularly for greater use of biomass); but sustainability criteria must be applied and may limit the scope for expanding this technology. Use of biomass as a heat source for pyrolisis, coupled with production of biochar (or charcoal) that can be applied to soil as a soil improvement and carbon retention agent, looks promising. In the transport sector, electricity or hydrogen appears to be more promising as long-term solutions. Intensive discussions with the relevant industry sectors should continue to try to establish optimum pathways for these transitions.

Taxation of oil products already produces major income for governments, and this may increase further when carbon permits are auctioned more extensively. Part of these proceeds should be directed to supporting the development of the new energy technologies that are required. Incentivizing the oil industry itself to play a larger part in the transition to the low-carbon economy by offering them tax concessions or capital allowances with respect to the necessary investments might also be considered.

Trade unions have an important part to play as well. Many of their members are in the front line in delivering energy-efficiency improvements and disseminating practical information and their potential contribution needs to be recognized and encouraged. Properly managed, the new forms of production should provide just as many employment opportunities as the older carbon-intensive modes of production, while maintaining good working conditions.

Conclusions

The world is facing a dual crisis from accelerating climate change and growing energy insecurity. Our prodigal use of oil is at the heart of the crisis. There is still a chance for the world to navigate its way out of the crisis successfully if vigorous action to reduce our dependence upon oil is put in hand urgently. But we need 'Action this Day'.

Note

1 This chapter is based on an opinion delivered by the European Economic and Social Committee to the European Parliament in January 2009. The facts and analyses underlying it are derived from well-known international sources, such as the Intergovernmental Panel on Climate Change (IPCC) and the International Energy Agency (IEA), and are too well known to need repetition here. The style is deliberately abrupt and staccato to concentrate attention on the enormity and urgency of the challenge and the wide range and scope of the actions needed to cope with it. The recommendations were drawn up for Europe, so some of the points in this chapter address the European situation. But, *ceteris paribus*, they can be applied equally to the situation of most other developed countries, and increasingly to the emerging economies of the developing world.

References

IEA (International Energy Agency) (2008) *World Energy Outlook 2008*, IEA, www.worldenergyoutlook.org/docs/weo2008/WEO2008_es_english.pdf.

IPCC (Intergovernmental Panel on Climate Change) (2007) *Climate Change 2007: Impacts, Adaptation and Vulnerability. Contribution of Working Group II to the Fourth Assessment Report of the Intergovernmental Panel on Climate Change*, Cambridge University Press, Cambridge, UK

4
The Nuclear Option

Elizabeth Dowdeswell

Introduction

Any analysis of the current state of geopolitics and economic development reveals a pervasive sense of insecurity about future energy supply and global climate disruption. How to meet our ever increasing energy needs and wants in a responsible and environmentally sustainable way is one of the most vexing social and technological conundrums facing the world today.

The most recent World Energy Outlook report of the International Energy Agency (IEA) describes the rise in energy demand as 'alarming' and paints a pessimistic picture of uncertainty in production capacity and soaring prices. The current demand of 85 million barrels a day is predicted to rise to 116 million barrels in 2030 with the increase in demand dominated by China and India. The IEA's reference case scenario also predicts that emissions of greenhouse gases will jump by 57 per cent between 2005 and 2030, with China overtaking the US in 2008 as the biggest emitter (IEA, 2008).

Coincident with growing energy demand, climate change is forcing us to face the inevitability of a carbon-constrained world. The analysis of the Intergovernmental Panel on Climate Change (IPCC) has left no doubt about the need for urgent collective policy action to redirect the world community to a more sustainable energy path. Their fourth assessment documents with clarity and greater certainty that 'warming of the climate system is unequivocal' and that most of the observed increase in temperatures is very likely due to an increase in anthropogenic greenhouse gas concentrations. Emissions continue to increase, with a substantial jump of 70 per cent between 1970 and 2004. Observational evidence and our understanding of the real and potential impacts are mounting.

A risk management strategy of mitigation and adaptation is being developed, albeit slowly, under the United Nations Framework Convention on Climate Change (UNFCCC). The discussion about the Bali Roadmap in December 2007 and 2008 drew attention to the challenge of developing a comprehensive global strategy that will engage all major emitters while recognizing the developing country imperatives of economic growth and poverty alleviation.

Thus, global climate disruption has opened the door for discussion about a potential nuclear renaissance to reduce our reliance on fossil fuels.

Nuclear power at a glance

According to the International Atomic Energy Agency (IAEA), currently there are 439 nuclear power plants in operation in 30 countries with a total installed capacity of 371 gigawatts, five plants in long-term shutdown and 34 under construction in 14 countries. While the US had the most operating units, half of the reactors now being built are in developing countries. At the end of 2006, nuclear provided about 15 per cent of total electricity production worldwide. Some 16 countries rely on nuclear energy to supply at least one quarter of their total electricity. France tops the list at 78 per cent.

In their latest report *Energy, Electricity and Nuclear Power Estimates for the Period up to 2030*, the IAEA (2006) projects two scenarios:

1. The low projection, which assumes only completion of all nuclear capacity currently under construction or firmly in the pipeline, would see nuclear capacity grow 25 per cent by 2030.
2. The high projection, which considers 'reasonable and promising' additional projects, forecasts a 93 per cent increase in capacity in the same period, representing an annual growth rate of 2.5 per cent (IAEA, 2006).

It is not surprising that there is speculation about a nuclear renaissance. Observers report that China plans a fivefold increase in nuclear generation by 2020, while India plans an eightfold increase by 2020. Serious consideration is being given to new build in the US, Canada, Russia, Argentina, Japan, South Africa, the UK and other European countries. The first new plant in Europe in many years is under construction in Finland. Countries as diverse as Algeria, Belarus, Egypt, Indonesia, Iran, Jordan, Libya, Nigeria, Thailand, Turkey, Vietnam and Yemen are among those examining the possibility of nuclear power generation as a contribution to realizing their development ambitions.

Nonetheless, there are several countries such as Austria, Australia, Italy, Denmark and Ireland that have declared a prohibition against nuclear development. Others, like Germany and Belgium, have announced their intention to phase out nuclear energy production.

The case for nuclear

In the search for low-carbon energy supply options, nuclear is under reconsideration as part of a diversified mix. There are those who advocate that nuclear is, at present, the only safe large-scale energy source available for base load power. Soothing corporate advocacy campaigns claim that nuclear is clean, affordable and reliable. Industry advocates have been joined by some environmentalists, including Stewart Brand, author of *The Whole Earth Catalogue*, Gaia guru James Lovelock, and Patrick Moore, formerly of Greenpeace. They claim that the risks from failing to reduce atmospheric carbon dioxide outweigh those associated with the use of nuclear power.

Research by the Nuclear Energy Agency (NEA) of the Organisation for Economic Co-operation and Development (OECD) and the International Atomic Energy Agency (IAEA) concludes that throughout the life cycle nuclear power emits low amounts of carbon dioxide, between 7g per kilowatt hour and 22g per kilowatt hour – an amount similar to the carbon dioxide emissions from wind power and considerably less from fossil-fuelled plants.

The oft-quoted interdisciplinary *Massachusetts Institute of Technology* (MIT) study on the future of nuclear power concluded that the nuclear option should be retained in a carbon emissions management strategy 'precisely because it is an important carbon free source of power' (Ansolabehere et al, 2003). The study focused on analysing the critical problems that would have to be overcome. A recent Inter-Academy Council report came to a similar conclusion that 'as a low carbon resource, nuclear power can continue to make a significant contribution to the world's energy portfolio in the future' (Inter-Academy Council, 2007). This panel also introduced the caveat that there were major concerns to be addressed.

For many years, the subject of nuclear power was peripheral to the international climate change negotiations. However, during 2007, the IPCC in its fourth assessment declared that nuclear power 'has the potential for an expanded role as a cost effective mitigation option' (IPCC, 2007) and identified constraining factors. The UNFCCC estimated that to achieve the necessary shift away from fossil fuels, the 'investment in nuclear generation would need to increase from $15 billion to $40 billion, including an additional $11 billion investment in developing countries'.

The UK has just concluded a consultation on the role of nuclear power within that government's energy strategy. Their consultation document is illustrative of increasing government initiatives to articulate the case for considering nuclear power in the context of tackling climate change and energy security issues. Their preliminary view is that new nuclear power stations could make an important contribution. And although it is the sovereign right of each country in the European Union to make its own energy supply decisions, the European Commission has recognized the need to have a platform for debate among researchers, industry and environment ministers.

With the aim of better informing the debate, several reports have been issued recently. Illustrative are the following three examples from 2007. A report from the Oxford Research Group, an independent non-governmental organization in the UK, suggests that claims about the impact nuclear power could have upon global carbon emissions should be reviewed and questions whether or not the security risks could be managed (Barnaby and Kemp, 2007).

Charles Ferguson addressed the risks and benefits of nuclear energy on behalf of the Council on Foreign Relations in the US. He also examines the validity and feasibility of a significant nuclear contribution to reducing global warming and the potential dangers associated with worldwide proliferation of this energy source (Ferguson, 2007). Third, The Ethical Funds Company of Canada prepared a report entitled *One Is Too Many* (Ethical Funds Company, 2007) for shareholders. It reviews the potential for revival of nuclear as a primary strategy to fight climate change and concludes that there are better options. As nuclear power contributes only about 2 per cent to total world energy needs (as opposed to that provided), dramatic and rapid increase in new build would be needed to make a significant contribution in climate change mitigation. Critics conclude that nuclear power does too little, too late, with unacceptable risk.

Challenges to a nuclear renaissance

The fact that employment of nuclear energy has not been fully embraced points to unanswered questions and interrelated challenges in its implementation. For some, it is a question of feasibility. Can nuclear expansion occur quickly enough and on a scale commensurate with the need? For others it is a question of economics. Comparative to other energy supply options, is it worth the investment? Will future development have to rely on massive government subsidies? Others cite the continuing quest for an acceptable approach to management of long-lived wastes and uncertainty about risks to human health and the environment. Whether nuclear power is dangerous remains a question for many. The fears and insecurities generated by Chernobyl and Three Mile Island have only intensified post-9/11. The threat of terrorism and the nuclear weapons ambitions of certain states have led to speculation that the international regime may not be robust enough to counter the inherent proliferation risks.

Proliferation

The link between an expansion of nuclear energy capacity and the potential spread of nuclear weapons is a continuing cause for concern. The authors of the MIT study probably reflect the views of many when they recommend that nuclear power should not expand unless the risk of proliferation is made acceptably small (Ansolabehere et al, 2003).

Security risks arise with the possible diversion or theft of materials such as stocks of stored plutonium from nuclear facilities with inadequate controls. Perhaps of even greater concern is the risk that an increasing number of nations could acquire sensitive technologies or materials that would enable them to develop a nuclear weapons capability. Fuel cycles that involve reprocessing of used fuel to separate out weapon-usable plutonium or the enrichment of uranium are specific examples. The number of countries with access to fissile materials and competent nuclear physicists and engineers who could design and fabricate nuclear weapons is likely to increase. Some of those countries are likely to be in politically unstable regions with differing security circumstances.

Post-9/11, a related concern has been the security of existing nuclear facilities against terrorist elements. The potential of targeting a facility or materials in transit in order to cause a major radioactive release, mass casualties, significant economic impact or simply confusion has mobilized many plants to improve physical protection and security operations. A scenario in which terrorist organizations could acquire materials to create crude nuclear devices such as a dirty bomb is everyone's nightmare.

The IAEA was created in the late 1950s to administer safeguards while promoting the peaceful applications of nuclear science and technology. A key instrument is the Treaty on Non-Proliferation of Nuclear Weapons, which entered into force in 1970 and has been ratified by nearly 190 countries (IAEA, 1970). The bargain that was struck was that non-nuclear weapons states would give up the development or acquisition of nuclear weapons in exchange for the 'inalienable right' to nuclear energy for peaceful purposes. The nuclear weapons states (the US, Russia, China, France and the UK) would pursue negotiations towards nuclear disarmament. Additional protocols have been developed to strengthen safeguards through inspection and detailed accounting processes. In parallel, a voluntary Nuclear Suppliers Group of 45 member countries seeks to contribute to non-proliferation of nuclear weapons through coordination, implementation of guidelines and control of nuclear related exports.

The regime is not without its difficulties and continues to be tested. Certain countries have chosen to remain outside the treaty, while others can develop their technologies and facilities and then opt out. Furthermore, private networks of technology suppliers circumvent the controls of the treaty. In order to ensure that the IAEA is a strong, independent verification organization, adequate financial resources are required. Some suggest that tighter controls are necessary to keep nuclear material stockpiles secure and that mechanisms need to be developed to restrict deployment of reprocessing and enrichment technologies. A proposed agreement that would allow India to acquire nuclear fuel and advanced reactors from the US in return for permitting inspection and safeguards is thought by some to strike at the very essence of the non-proliferation regime. And, of course, nations need to make progress on nuclear disarmament.

Improving proliferation resistance is also a key objective in developing the next generation of technologies. Collaborative efforts to develop advanced closed-fuel cycle systems, more efficient use of uranium and thorium and minimization of waste are being undertaken by the ten nations and the European Union within the Generation IV International Forum and the International Project on Innovative Nuclear Reactor and Fuel Cycles of the IAEA. Proposals involving fuel leasing and take-back are surfacing. The Global Nuclear Energy Partnership seeks to guarantee supplies of nuclear fuel and services to certain countries eliminating the need for them to develop indigenous fuel-cycle capabilities. The concept of 'insure to assure' has recently been proposed by a joint team from the Wharton Business School and Harvard's Kennedy

School. The proposal envisages a partnership between financial industries and governments to create the world's first international nuclear fuel insurance fund that would discourage the spread of enrichment facilities.

So, while some nuclear power proponents suggest that nuclear proliferation and terrorism risks are readily managed, others allege weaknesses in the system and warn that there is no room for complacency.

The economics

Conventional wisdom holds that while there are substantial capital costs for the construction of nuclear power plants, the actual operating costs compare favourably with other sources. Although the true cost of nuclear power is difficult to determine, the 2003 MIT study estimates a cost of 6.7 US cents per kilowatt hour and 2004 University of Chicago studies claim 6.2 US cents per kilowatt hour (University of Chicago, 2004). New technologies, as yet unproven, and equitable carbon pricing policies could make nuclear an attractive economic option.

Proponents and critics alike acknowledge that economics is a major factor influencing the growth of nuclear energy. Critics claim that nuclear power will only attract the significant private investment required if governments guarantee profits and markets and assume major risks and liabilities for cost overruns, fuel and waste disposal costs, decommissioning and accidents. Their view is that nuclear power plants are not economic without direct and hidden subsidies. Virtually all of the nuclear power plants in operation today were developed by government-owned monopolies, with investors insulated from many risks that are borne by taxpayers and consumers.

Investors are cautious about taking on the financial risks associated with capital-intensive projects prone to cost overruns. The long lead times, perhaps a decade from design, through government approvals and construction to completion and regulatory uncertainty, have been significant barriers. Even though worldwide operational performance has improved (the IAEA records an average unit capacity factor of 83 per cent in the period of 2003 to 2005) and average operating costs have decreased in countries such as the US, greater incentives will be necessary.

These range from reform of the licensing process, perhaps by combining construction and operating licences in one application, limiting redesign requirements, and improving performance in operations that have been marked by unplanned and costly shutdowns. Design of simpler reactors and extension of licensed lifetimes would improve the economic case.

There is a worldwide trend towards privatization of electricity generation. The UK's consultation document on the future of nuclear power states: 'It would be for private-sector energy companies to propose and fund the construction and operation of any new nuclear power stations – including full costs of decommissioning and full share of waste management costs' (UK DTI, 2007). The consultation document also recognizes that governments can play an important role in reducing regulatory uncertainty. Most analysts argue that stimulating additional investment will require policy initiatives, such as pricing of carbon, setting up an emissions trading system, and providing tax credits and subsidies in order to demonstrate cost and regulatory feasibility.

Waste management

When used nuclear fuel is removed from the reactor it is highly radioactive, and requires proper shielding and careful handling to protect humans and the environment. Although radioactivity decreases with time, nuclear fuel remains a potential health, safety and security hazard for thousands of years – some say indefinitely. The world has to deal with a legacy of tens of thousands of tonnes of highly radioactive waste.

The nuclear industry broadly supports deep geological disposal as the preferred approach to managing the waste. Notwithstanding considerable research about the science, technology and engineering of possible storage and repository approaches over many years, the task of implementation has proven challenging. While repositories are in operation for disposal of low- and intermediate-level wastes in certain countries, no country has yet begun to store used fuel from commercial plants in permanent repositories. Some nations have declared, or even legislated, that a deep geological repository that isolates and contains used nuclear fuel is the appropriate approach. In addition others reprocess waste, partially to reduce waste volumes. They also provide waste reprocessing services to other countries. Transmutation technologies are the subject of extensive research in France.

While Finland has begun detailed rock characterization studies for its underground repository at Olkiluoto, and Sweden is well on the path to choosing between two possible sites for disposal, the controversial chosen site in the US – Yucca Mountain – has been delayed on technical grounds and because of local opposition. These delays have resulted in lawsuits and financial penalties. Several countries (the UK, France and Canada) have just completed reviews of the waste management options, while others have not yet decided on a path forward. In the interim, usually spent fuel is immediately stored in water pools and later, after cooling for five to ten years, moved to interim storage in dry casks on site. Internationally, there have been some discussions about regional and international repositories to ease the cost burden on smaller countries.

Future technological developments might change the balance of risks. Advanced closed-fuel cycles separate or partition plutonium and other actinides, transmuting them into shorter-lived and more benign species. The goal is to reduce waste heat, increase the storage capacity of the repository and shorten the required waste isolation time. Debate continues as to whether or not these potential benefits outweigh the incremental costs, short-term safety and environmental risks.

Over the past decade a number of national waste management programmes have had to be reigned in and rethought, put on hold, or even abandoned in the face of public opposition and activist electorates. Radioactive waste decisions, once considered the exclusive purview of governments and the nuclear industry, are now clearly in the public domain. The analyses in Canada (National Waste Management Organization, 2005) and the UK were based on legacy wastes and explicitly say that the creation of new wastes involves ethical considerations that should be part of a wider discussion about the future of nuclear. Technically, new waste could be accommodated with an increase in the size and cost of the facility.

The time dimension of managing nuclear waste is a key ethical issue. Effectively, we are being asked to develop public policy that would require implementation over a period longer than recorded history. Given the longevity of the hazard, it is imperative that obligations to future generations are considered explicitly. Managing nuclear waste requires a great margin of safety, as damage would be multigenerational and irreversible. While the probability of an event may be low, the consequences could be high. Technically, the waste problem may be manageable. But any approach has to be, and be perceived to be, robust under all scenarios.

Human and environmental health and safety

Nuclear power generation is not without human health and environmental risk. In fact, this is an underlying theme in any consideration of the issues of proliferation, economics and waste management, and the subsequent ambivalence about embracing nuclear unequivocally as an appropriate response to climate change.

Any full life-cycle analysis would point to potential emissions from reactor construction and mining as well as from the actual nuclear reactor operation. There are radioactive tailings, leftover sludge, contamination of groundwater and the surrounding environments to be dealt with. Reactors use huge quantities of water. They are also potential sources of routine and accidental releases of radio nuclides and other toxic substances. And, of course, radioactive wastes are themselves highly toxic and include elements with very long half lives. Direct exposure is fatal. Indirect exposure, through seepage into groundwater, for example, can lead to life-threatening illness.

Physical, chemical and biological stresses on wetlands, waterways, land, groundwater and air sheds can result from activities as diverse as the building of necessary infrastructure, transportation, mining, fuel fabrication, transmission of electricity and, ultimately, decommissioning and management of waste. Furthermore, the effects may be local and widespread, cumulative and intergenerational. Another consideration is that nuclear power generation is not entirely carbon neutral. The estimates of greenhouse gases produced vary widely depending upon the assumptions made about factors such as the quality of the uranium ore used for fuel, waste management and decommissioning requirements.

While modern reactors can achieve a very low risk of serious accidents, the 1979 and 1986 reactor accidents at Three Mile Island and Chernobyl, respectively, have had a lasting impact upon public confidence. Undoubtedly, concern about safe and secure plant operations, ageing equipment and materials and transportation of nuclear materials have grown since 9/11. Due to the fact that there has, of yet, been no experience with waste repositories for used nuclear fuel, uncertainty remains about their safety.

Arguably, through long-term monitoring of environmental impacts, rigorous environmental assessment processes and regulatory regimes, we are identifying and mitigating risk to the environment. There is opportunity to assess and mitigate impacts so that environmental integrity is maintained over the long term.

A final word

There is perhaps no more quintessential public policy challenge facing society today than the responsible management of nuclear energy. The challenge is to be able to benefit from technology and to reduce risk, while also respecting the values of our citizens. Yet, research shows that public trust in governments and major institutions is eroding while society's expectations to be involved in decision-making have become more intense and sophisticated.

Public support will be vital. In order to obtain a 'social licence' to proceed, it is incumbent upon decision-makers to understand the terms and conditions that would make nuclear power acceptable to society and to respect these factors in its design and implementation. Members of the public have a right to be engaged in discussions about matters that affect their lives fundamentally. Simply seeking the best technical approach is not enough. While technical and scientific specialists will articulate and help us to understand the nature of the risk, the technical adequacy of options, environmental impacts and economic feasibility, ultimately it is society at large who will determine which risks it is prepared to accept.

Values and ethics are absolutely central. Since ethical questions rarely have unambiguous or definitive answers, past attempts to solve them through technical arguments alone have not been satisfactory. As in any complex issue, trade-offs among competing objectives are going to be inevitable. In a democratic society, the inclusiveness and integrity of the process by which decisions are taken will be key to the success and acceptance of such a large socio-scientific project. Only a process that considers diverse views deliberately and transparently will be considered trustworthy of protecting the public interest.

Decisions will have to be taken in a dynamic and adaptive rather than static manner. Technological and societal changes will be inevitable. The public gravitates towards a precautionary approach instinctively because they are humble about the state of our current knowledge and uncertainties over time, optimistic about the future and respectful of decisions made today for future generations. At every stage, earning and retaining trust and confidence will be essential.

References

Ansolabehere, S., J. Deutch, M. Driscoll, P.E. Gray, J.P. Holdren, P.L. Joskow, R.K. Lester, E.J. Moniz, N.E. Todreas (2003) *The Future of Nuclear Power: An Interdisciplinary MIT Study*, MIT, Cambridge, MA, web.mit.edu/nuclearpower

Barnaby, F. and J. Kemp (2007) *Too Hot To Handle: The Future of Civil Nuclear Power*, www.oxfordresearchgroup.org.uk/publications/briefing_papers

Ethical Funds Company (2007) *One Is Too Many: Considering Nuclear in a Time of Climate Change*, www.ethicalfunds.com/SiteCollectionDocuments/docs/Considering_Nuclear_Final

Ferguson, C. (2007) *Nuclear Energy: Balancing Benefits and Risks*, Council on Foreign Relations, New York, NY, www.cfr.org/content/publications/attachments/NuclearEnergyCSR28

IAEA (International Atomic Energy Agency) (1970) *The Treaty on Non-Proliferation of Nuclear Weapons*, www.iaea.org/Publications/Documents/Infcircs/Others/infcirc140.pdf

IAEA (2006) *Energy, Electricity and Nuclear Power Estimates for the Period up to 2003*, www-pub.iaea.org/MTCD/publications/PDF/RDS1-26_web.pdf

IEA (International Energy Agency) (2008) *World Energy Outlook 2008*, www.worldenergyoutlook.org/2008.asp

Inter-Academy Council (2007) *Lighting the Way*, www.interacademycouncil.net

IPCC (Intergovernmental Panel on Climate Change) (2007) *IPCC Fourth Assessment Report*, University of Cambridge, Cambridge, www.ipcc.ch/ipccreports/ar4-syr.htm

National Waste Management Organization (2005) *Nuclear Waste Management Organization, Choosing A Way, Forward: The Future Management of Canada's Used Nuclear Fuel*, NWMO, Toronto, www.nwmo.ca/Default.aspx?DN=b48e625b-d71b-4e83-a859-83e4fccb08e6

UK DTI (Department of Trade and Industry) (2007) *The Future of Nuclear Power: The Role of Nuclear Power in a Low Carbon UK Economy*, UK DTI, www.berr.gov.uk/files/file39197.pdf

University of Chicago (2004) *The Economic Future of Nuclear Power*, www.anl.gov/Special_Reports/NuclEconAug04.pdf

5
Bioenergy: Neither Golden Solution nor Prescription for Disaster

Jürgen Maier

Introduction

There is probably no other energy issue currently being discussed with so much emotion and so little facts as bioenergy. For some years, biofuels have been promoted by governments and legislators in Europe and the US with only few questions being asked about side effects and their actual sustainability. Biofuels rapidly were perceived as the golden solution to climate change and energy insecurity. However, questions are now being asked by a range of non-governmental organizations (NGOs), activists and journalists, often with little consideration for the potential benefits of biofuels or bioenergy, in general. For many NGOs and activists, biofuels almost overnight became the single most important threat to food security, biodiversity and the survival of the rainforests. UN Special Rapporteur on the Right to Food Jean Ziegler expressed it in his usually drastic way:

> *If there are not conscious efforts to ensure that producing biofuels does not bring greater hunger in its wake, then the poor and hungry will be the victims of these new fuels. For Fidel Castro the case is equally clear: it is a 'sinister idea to transform food into fuel'.* (Ziegler, 2007, p9)

This state of the debate is unfortunate. We urgently need a rational debate about how the immense potential of modern bioenergy can be utilized for a sustainable energy future. If we are to move to a sustainable energy future, we have to phase out fossil fuels and nuclear energy, currently used by the 4 billion people that have access to modern forms of energy. We also have to phase out traditional biomass currently used by the 2 billion people who have no access to modern forms of energy. This daunting task is impossible with solar, wind, small hydro and geothermal energy alone, no matter how much energy efficiency is applied. There is no alternative to a massive utilization of bioenergy in any sustainable energy scenario for the future.

Any responsible energy policy therefore needs to integrate a number of conflicting issues: the prevention of dangerous climate change by a rapid reduction of greenhouse gas emissions, economic development and poverty alleviation in developing countries; the preservation of biodiversity, food security and the abolition of hunger; as well as a number of other issues. These conflicts are mirrored in the bioenergy debate like nowhere else, but it is not an impossible task.

This chapter will look at the role of biofuels and bioenergy in the context of climate and energy, as well as agriculture and poverty alleviation.

Climate and energy scenarios

The reports from the Intergovernmental Panel on Climate Change (IPCC) in 2007 – as well as numerous other climate studies – are alarming. Climate change is taking place much faster than the scientific community expected only a few years ago. At the same time, the world's carbon dioxide emissions are still rising and in many countries the growth in emissions is accelerating. If we are to stop dangerous climate change, global emissions must peak no later than 2015 and be reduced by at least 50 per cent by 2050 (IPCC, 2007). However, the world is far from being on track for such a scenario. The message is clear: there is an urgent need to act now to speed the global energy revolution away from fossil fuels.

Alternative energy and emission scenarios aimed at phasing out nuclear and fossil energies have consistently put their emphasis on two pillars: massive expansion of energy efficiency and of renewable energies. Two recently released studies are from Greenpeace and the European Renewable Energy Council (EREC) (Greenpeace and EREC, 2007) and from the German government's Advisory Council on Global Change (WBGU, 2007). Their scenarios, projecting future energy, are based on the need to cut global greenhouse gas emissions consistent with the IPCC findings. They also assume the same global primary energy consumption as today, with 50 per cent coming from energies produced from fossil fuels. Such a scenario would have only a slim chance of keeping the global average temperature rise below 2°C. Both scenarios call for a massive energy-efficiency revolution and a massive increase in renewable energy. The Greenpeace/EREC scenario sees a tenfold increase of global bioenergy use for electricity production by 2050, 60 per cent coming from energy crops. The WBGU study calls for an increase in bioenergy for electricity production from 1.5 per cent globally to 30 per cent by 2025, and wind from 1 to 20 per cent by 2025.

The question now is: what is the most efficient use for bioenergy? No other renewable energy source can be used in such manifold ways. A key question here is the 'greenhouse gas emissions balance' – carbon dioxide released from its combustion must be 'captured' by plants grown in the next productive cycle. This is not always necessarily the case. Bioenergy sources generated by a massive input of fossil fuel sources (fertilizers, processing, transport, etc.) can have a greenhouse gas emissions balance that is hardly any better than fossil fuels. The sustainability of an energy source can, in the end, only be determined by a life-cycle assessment of the entire production and utilization chain of each energy source.

As soon as climate protection (meaning the reduction of fossil fuel combustion) is recognized as a key driver for the utilization of bioenergy, priority must then be given to those forms of biomass that reduce the largest amounts of carbon dioxide emissions and are the most energy efficient. Processed into liquid fuels, bioenergy as indicated in life-cycle assessments is five to ten times less energy efficient than biogas, wood or woody biomass.

In the immediate future this means that, in most countries, the use of biomass in the electricity and heating sector will achieve the highest greenhouse gas emissions substitution potential by replacing coal, the most greenhouse gas-intensive fossil fuel. In the electricity and heat sector, agricultural residues and biomass waste, as well as unprocessed biomass from the forestry and agricultural sectors, are the inputs that are most suitable. Biomass, in contrast to the intermittently available renewable energy sources such as wind and sun, is available at all times. In combination with wind and sun, biomass can be crucial to maximize the potential of all three renewable energy sources as a standard base load energy source.

However, greenhouse gas reduction is only one driver for biofuels. Energy security, the dependence upon imported petroleum with all its economic and political problems, is a second powerful driver for biofuels, often overriding climate concerns. The transport sector of any modern economy is strongly dependent upon petroleum. Therefore, the petroleum and automobile industries put particular emphasis on using biomass for the production of liquid fuels. However, due to processing and refining, the use of biomass results in greater reductions in carbon dioxide emissions. Clearly, the development of sustainable energy policies also has to address the transport sector.

Biodiesel and bioethanol

Ambitious biofuels targets have been agreed in major automobile markets. In the US, a goal of 35 billion gallons of biofuels, replacing approximately one fifth of oil-based transport fuels, is proposed by 2017. Current production is just 4.2 billion gallons. The European Union (EU) Biofuels Directive demands that 5.75 per cent of European fuel comes from biomass by 2010, increasing to 10 per cent by 2020. Even more ambitious targets are being discussed (Doornbosch and Steenblik, 2007).

Bioethanol is the most widely used biofuel for transport and accounts for more than 94 per cent of global biofuel production, the rest being mostly biodiesel from various oil plants such as palm oil, soy and rapeseed (IEA, 2008, p161). About 60 per cent of the world's bioethanol comes from sugarcane and 40 per cent from other crops. Germany is the market leader in biodiesel where bioethanol is insignificant. With sales at approximately 1.5 million tonnes, biodiesel currently supplies more than 4 per cent of the demand for diesel. The planned increase to 2 million tonnes would allow biodiesel to cover approximately 6 per cent of the current demand. Germany currently has about 750,000ha for biodiesel production; the land available for biodiesel production cannot, however, be extended to more than 1 million hectares. The international trade in biodiesel is only beginning.

Another question is whether plant oils have to be refined into diesel at all. Vehicles can also run on pure plant oil. The energy (and agricultural) balance of pure plant oil from perennial plants such as Jatropha on marginal land is much better, particularly when used for rural electrification in diesel generators.

The biofuel of choice is clearly Brazilian ethanol made from sugarcane, since both in terms of production costs and greenhouse gas balance it tops most other biofuels currently being used. Brazil is by far the largest bioethanol producer and also the largest exporter supplying about half of the global market. However, despite the rhetoric of the Brazilian government about massive exports, Brazil's sugarcane producers themselves put their increased export potential in the near future, at 20 per cent over present levels (Carvalho, 2007).

Brazil produces ethanol first and foremost for its own domestic market. Europe and the US are protecting their own biofuel industries with heavy tariffs and subsidies against cheaper imports. At the same time, Brazilian bioethanol is much more competitive than bioethanol from corn/maize made in the US, for instance, and its energy and greenhouse gas balance is also much better. The question, then, is why protect economically and environmentally inefficient American or European producers from international competition?

Reorganizing mobility

Neither bioethanol nor biodiesel are the best option to reduce emissions and oil dependence in the transport sector. It is impossible to supply a major share of today's global market of liquid transport fuels based on bioethanol and biodiesel. Not only is the production of these fuels too inefficient, but so is the transport system itself. There is no way to get out of petroleum's grip without addressing the way in which mobility is organized today. It should be pointed out that the EU biofuels target of 10 per cent by 2020 can be achieved by reducing the consumption of fossil fuels in the transport sector and, thus, automatically increasing the share of biofuels without actually increasing their absolute amount.

This requires a modal shift in the transport sector. The automobile-centred transportation system dominant today is energy inefficient and cannot be extended to the whole world. It is not possible to produce the enormous amount of fuel needed for this transport system in a sustainable manner. The removal of the many counterproductive subsidies for aviation and road transport is a prerequisite for a reduction of absolute transport demand by discouraging excessively dispersed production cycles and settlement patterns. Modern and attractive public transport systems, including massive expansion of railways, for both passenger and freight transport, are the cornerstones of a sustainable transport system of the future.

Taking all that into account, there will still be a need for biofuels for motorized vehicles. Strict fuel-efficiency criteria and policies to take gas guzzlers off the market, increasingly implemented in places such as California and China, will be inevitable. This will have much more impact for reducing the petroleum consumption of the transport sector than any shift to alternative fuels. To put it bluntly: 1000 people using public transport powered by conventional electricity are emitting much less greenhouses gases than 1000 people driving biofuel-powered sports utility vehicles (SUVs).

Biogas: The biofuel of choice

For alternative fuels, we need to look beyond ethanol or biodiesel. More and more studies show that biogas is more efficient than bioethanol produced from sugarcane (Brazil) and even more efficient than bioethanol from maize or sugar beets. Biogas (biomethane) is even more impressive as a fuel when net energy yields per hectare are compared. The gross biofuel yield from 1ha of agricultural land is 4977 litres of fuel equivalent for biogas, 4179 litres for ethanol from Brazilian sugarcane, 4054 litres for ethanol from sugar beet, 3907 litres for biomass to liquid (BTL, so-called second-generation liquid biofuels), a lousy 1660 litres for bioethanol from corn, and 1408 litres for biodiesel from rapeseed.

The picture is even more pronounced when you look at the net energy balance – when you take into account the amount of energy necessary to obtain the energy stored in bioenergy. Today, biogas from silage maize, Sudan grass and other productive energy plants already produces a net energy yield of 42,000 to 62,000 kilowatt hours (kWh) per hectare. In contrast, BTL can only produce around 33,600kWh. The energy-intensive process of distilling bioethanol from whatever source generally produces an even less impressive net energy yield, depending upon the kind of energy input. In terms of carbon dioxide saving costs, biogas is the most favourable of all biofuels. In addition, second-generation biofuels are produced by transforming biomass in a high temperature (energy-intensive) chemical industrial process into liquid fuels. This does not allow for the recycling of plant materials as fertilizers, thus requiring chemical fertilizers with a corresponding additional energy input. The residues from second-generation biofuels are essentially industrial waste. In contrast, biogas production takes carbon out of the agricultural cycle and the residues with all the nutrients can be recycled to the agricultural system. In principle, biogas production is compatible with organic agriculture. Compared to bioethanol and biodiesel as well as to the second-generation liquid biofuels, biogas is more efficient. The biofuel of choice should therefore be biomethane/biogas, and the biofuels quota should be primarily met with biogas.

Biogas for transport is a technology already available worldwide since biomethane is physically identical with natural gas. Today there are more than 6 million natural gas/compressed natural gas (CNG) vehicles already operating globally. Biogas production is also an interesting component for sanitation systems. Coupling sanitation systems with biogas production can provide the necessary water for biogas production (important in arid regions) and make an important attractive link between energy and sanitation.

Electric vehicles

More attention has recently been placed on electric vehicles. Depending upon the way in which the electricity to run them is produced; they may well be in the longer run the most promising option for motorized vehicles. However, in a future electricity mix a considerable part will be produced from biomass, so even for electric vehicles there will be a substantial bioenergy component. A shift to both biogas and electric vehicles will have enormous economic and political consequences because such a shift will take the powerful oil companies completely out of fuel production and perhaps even out of fuel distribution. It is also not the optimal choice of most automobile companies who have all invested in other options. The biogas option would mean a dramatically increased role of farmers and natural gas companies, at the expense of the oil companies. The shift to electric vehicles would leave no role for oil companies at all. Don't be surprised if there is a substantial resistance from this sector against these options.

Agriculture and poverty alleviation

One of the biggest debates around bioenergy-sector development is the concern over food security with fuel crops replacing food crops. More than 800 million people today go hungry – not because the world does not produce enough food, but because they are too poor to buy it. Meanwhile, the rich in industrial and developing nations have the buying power to make sure that sizeable amounts of land are used to cultivate animal feed, and then buying it 'processed' as meat and dairy products. This situation is transferable to bioenergy crops.

Maize, soybeans, sugarcane and other crops can serve as biofuel feedstock, as animal feeds or as food for human consumption. As energy prices increase, producers exploring multiple markets may respond by shifting existing output of maize or soy from food/feed to fuel use. Or they may shift production from food/feed to non-food plants (for instance Jatropha). In both cases, increased prices and reduced supplies may have adverse effects on the poor and hungry, particularly in net food-importing countries. The actual situation, however, is somewhat more complex.

Recent increases in ethanol production from corn in the US led to a 400 per cent rise in the price of tortillas in Mexico, the staple food for the poor, since cheap corn imports from the US disappeared from Mexican markets. However, it is important to point out that the problem did not start with bioenergy, but with forced agricultural market liberalization and the integration of agriculture within global markets. Countries became net food-importing countries and have lost their food sovereignty resulting from this, not because of bioenergy. Only ten years ago, before the North American Free Trade Agreement (NAFTA), there was no US corn on the Mexican market. Cheap subsidized US corn exports destroyed the livelihood of many Mexican farmers. The small-scale agricultural sector in many developing countries has been destroyed or negatively affected by the dumping of subsidized agricultural products from European countries and the US. Small-scale farmers were and continue to be driven out of business, leading, in turn, to the migration of rural communities into urban areas. The dumping rates into other markets for some agricultural products from the EU to developing countries includes wheat (29 per cent), milk (42 per cent), sugar (56 per cent), poultry (26 per cent) and rice (26 per cent).

If bioenergy development in the EU and the US leads to reduced surpluses and less agricultural exports to developing countries, then this is certainly good news for farmers in developing countries. It is not necessarily good news for poor people who are not farmers but have to buy their food. Agriculture exports from the EU and the US have kept world market prices of a number of crops, such as maize, artificially low. When they are removed, prices for these crops will go up. In parts of the world such as Southern Africa, which is exacerbated by droughts and climate change, the effect may be high. Countries that import food are faced with price increases resulting from higher demand for certain crops. At the same time, food production in these countries becomes economically more viable and attractive. If managed well, bioenergy could promote something akin to an agricultural 'renaissance' in some developing countries where bioenergy can be produced profitably and also stimulate rural economic growth. Bioenergy can create jobs and protect incomes and livelihoods if it is decentralized and processed regionally, so that the added value stays in the region. Decentralized production of bioenergy holds a lot of potential, particularly in the case of raw materials such as Jatropha, castor oil plants and biogas, which can be converted into energy through simple technologies that can be easily installed on site and, as in the case with Jatropha, can be grown on marginal lands or through intercropping. The amount of arable land that is currently left idle because 'farm prices' are too low and that would come back into production with substantial increases of 'farm prices' can only be guessed, but it should not be underestimated.

The opposing model is a focus on liquid biofuels for export, which usually requires large areas for monocultures and a much higher grade of technology. Experiences with cash crops (including feedstock) in developing countries show that highly technical industrial cultivation methods need comparably large cultivation areas. In this case the land required will most probably be controlled by large landowners or companies. Large production units can also create jobs; but they usually simultaneously displace existing structures and thus create conflicts with access to land and water as well as the requirements of diversified agriculture driven by family businesses, co-operatives and rural communities aiming at supplying food and income for the local population, thus creating much more social inequity and poverty. The greater the commercial interest becomes, the greater the pressure from large-scale industry will be – akin to soy cultivation in Brazil. This could be the same in the case of bioenergy since production is linked to the land and there is no 'free' land. However, while we can probably make sure by regulation that bioenergy development will not increase social problems in developing countries with a large cash crop sector, it would at the same time be unrealistic to expect bioenergy production to meet higher standards than the rest of a country's agriculture.

It is, indeed, imperative to look at the place of bioenergy crops in agricultural markets as a whole in order to assess the impact that bioenergy production will have. The image of large-scale monocultures destroying Southern livelihoods and ecosystems for the production of some luxury items for export to European and other rich markets is far more appropriate for animal feed than for biofuels. Brazil's annual feed exports are about 35 million tonnes (up from 15 million tonnes in 1999), almost exclusively soya, and this represents about 55 per cent of Brazil's total agricultural exports. The value of these animal feed exports is about US$7.9 billion; 18.5 million hectares are under soy cultivation in Brazil. Almost half of Brazil's soya production is exported. Germany alone imports so much soy from Brazil that you can literally say 2 million hectares in Brazil are allocated for German meat consumers, a figure that corresponds to 10 per cent of German agricultural land. Despite these enormous animal feed imports, 85 per cent of German agricultural land is devoted to meat and dairy production. The world can feed every human being, as well as provide enormous amounts of bioenergy, but probably not if today's excessive consumption of cheap meat is expanded.

Compared to these massive figures, sugarcane production for ethanol in Brazil occupies currently 6 million hectares, and the potential for an increased 20 per cent export potential therefore represents 1.2 million hectares. This clearly shows that if there is an international market driving deforestation and unsustainable agricultural practices in Brazil, then it is animal feed and the demand for meat, and not biofuel production, which are responsible. Put simply, 1 million hectares in Brazil are occupied for ethanol exports, and 9.5 million hectares for soy exports. The Brazilian ethanol market is driven by domestic demand; the Brazilian soy market is essentially driven by, and for, the world markets.

The situation in Africa is clearly different. Arable land is scarce and mainly used not for cash crops for export but to feed the population, while cattle are grazing mainly on marginal grasslands. Bioenergy plants such as Jatropha, the Moringa tree and the indigenous plums in the region are suitable for marginal and degraded land, and in this case do not compete with food production; in addition, they may also have a beneficial effect through stabilizing degraded soil.

Apart from the social problems of large-scale agriculture, there are also serious environmental problems. A focus on cash crops for export often means that sooner or later high-yield varieties with a corresponding need for soil, irrigation, fertilizers and pesticides will be planted, often subsidized. Under these conditions, positive energy and greenhouse gas balances are difficult to attain and there will be negative environmental effects on, for instance, biodiversity. The use of genetically modified organisms in the production of industrial bioenergy raises many unresolved questions about risks (mostly environmental) and side effects (mostly health). The technology involved is also linked to the centralized control by a few large companies over seeds and plant varieties.

Another important issue is water availability. Many regions of the world are water scarce, and this will get worse with climate change. If biofuel crops are grown for export with irrigation on arable land, they will potentially not only compete with food but also for water, and that would be a development and environmental disaster. Agriculture already uses more than 50 per cent of all available water in many developing countries. It is imperative that among the diversity of bioenergy crops, those that are selected do not need excessive irrigation. It may well be that the limiting factor for bioenergy turns out not to be the availability of land, but the availability of water.

Unlike other agricultural markets, there is now a strong call from many important players to put sustainability standards for biofuels in place to ensure that the cultivation and export of bioenergy sources does not jeopardize a country's food supply. Governments must make sure that the production of enough food for the people gets clear priority. A regulatory framework is necessary to ensure food sovereignty and the right to food. This right is defined by the International Covenant on Economic, Social and Cultural Rights as follows:

> *Right to adequate food is a human right, inherent in all people, to have regular, permanent and unrestricted access, either directly or by means of financial purchases, to quantitatively and qualitatively adequate and sufficient food corresponding to the cultural traditions of people to which the consumer belongs, and which ensures a physical and mental, individual and collective fulfilling and dignified life free of fear.* (International Covenant on Economic, Social and Cultural Rights, 2002)

Policies to link biofuel incentives to some degree of mandatory sustainability certification are under way in The Netherlands, Germany and the UK. They are seen as pioneering a joint EU scheme that the commission is expected to work out in the near future. However, there much uncertainty about the practical side of how this can be implemented, as well as the extent of the sustainability criteria to be applied in such schemes. Social criteria are generally much harder to place into these schemes than environmental criteria. The UK and Dutch schemes are based on reports that the companies involved are required to file about how they implement sustainability criteria. However, companies can select their own 'sustainability criteria', and there are no sanctions if the reports are seen as incomplete. Indirect effects are not covered in either scheme.

It is important to note that we are talking about a work in progress here. The growing criticism of biofuel monocultures in developing countries will clearly have an impact upon the readiness of political will to legislate schemes that are not merely greenwashing exercises, including a certain determination to defend such schemes against possible producer country interference through the World Trade Organization (WTO).

Another way is voluntary sustainability certification that is modelled after the Forest Stewardship Council (FSC), a private body composed of timber industry, trade unions and NGOs. Multi-stakeholder dialogues such as the Roundtable on Responsible Soy (RTRS), the Roundtable on Sustainable Palm Oil (RSPO) or, more comprehensively, the Roundtable on Sustainable Biofuels (RSB) are trying to find common ground – certainly not an easy task but a necessary exercise. RTRS and RSPO are trying to define sustainability beyond the biofuel sector. There is a lot of criticism of these processes from NGOs and activists, claiming that they are not tough enough and highlighting the obvious potential for greenwashing. These roundtables are not the only game in town. As long as governments are not using such voluntary stakeholder processes to delay or reject mandatory regulation, their potential should neither be overestimated nor should they be dismissed as empty talk shops. The road to a more sustainable global agriculture – including the bioenergy sector, but not limited to it – is long and will be paved with many more multi-stakeholder dialogues, NGO actions, industry pioneers, government regulations, as well as international agreements. Sometimes they will contradict each other; but in the end it will be this mix and not just a single component that will lead to effective action addressing the problem.

Conclusions

The world is searching for a new energy paradigm – indeed, a new development paradigm. We know that this will have to be based on renewable energies and on maximum energy and resource efficiency. Nobody knows what the details of this new paradigm are going to look like; but we understand that business as usual is not an option. The way to these new paradigms will be littered with errors, mistakes and misjudgements – but they are necessary and inevitable to find the best solutions. The current state of affairs in the development of bioenergy reflects this. Many bioenergy options currently being pursued are neither sustainable nor economically viable. This is no surprise. The bioenergy sector in 20 years will look very different from what it is now, just as the wind energy sector in the last 20 years has seen enormous productivity and efficiency gains. Calls to stop bioenergy production or for a 'moratorium' on bioenergy would be counterproductive – we need better regulation and accelerated technology development to overcome wrong developments in the bioenergy sector.

In the face of limited resources and a limited resource base, efficiency needs to be – and will be – the key principle. Sufficiency is also an important issue; but this leads beyond the reach of this chapter. Bioenergy – and renewables, in general – cannot be just seen as a new energy input for the existing centralized energy and transportation system. A look at a sustainable and efficient utilization of bioenergy therefore needs to ask a few more questions than just which technology and which crop is the best. It also shows that we cannot look at a sustainable bioenergy future – or energy future, in general – without addressing some more comprehensive questions about how societies and economies should be structured to reduce the structural demand for energy input, while at the same time providing a comparable output of energy services. In the end, what we need or want is mobility, a warm home or similar things, not a sports utility vehicle or an electric heater.

References

Carvalho, P. E. (2007) 'Statement at Brazilian Sugarcane Growers Association', Press conference in São Paulo, Brazil, 21 May 2007, as reported by AFP

Doornbosch, R. and R. Steenblik (2007) *Biofuels: Is the Cure Worse than the Disease?*, OECD, Paris, www.oecd.org/dataoecd/15/46/39348696.pdf

Greenpeace and EREC (European Renewable Energy Council (2007) *Energy Revolution: A Sustainable World Energy Outlook*, January, www.erec.org/fileadmin/erec_docs/Documents/Publications/energy_revolution.pdf

ICESCR (International Covenant on Economic, Social and Cultural Rights) (2002) *Guiding Principle: Right to Food,* ICESCR, United Nations

IEA (International Energy Authority) (2008) *World Energy Outlook 2008*, IEA, Paris, p161

IPCC (Intergovernmental Panel on Climate Change) (2007) *Synthesis Report*, www.ipcc.ch

WBGU (German Advisory Council on Global Change) (2007) *New Impetus for Climate Policy: Making the Most of Germany's Dual Presidency*, WBGU, Berlin, www.wbgu.de/wbgu_pp2007_presse_engl.html

Zeigler, J. (2007) 'Report of the Special Rapporteur on Right to Food to the General Assembly', in *UN General Assembly, Doc.A/62/289*, 22 August 2007, UN, New York, NY, p9, www.righttofood.org/new/PDF/A62289.pdf

6
Localized Energy Conflicts in the Oil Sector

Nnimmo Bassey

Introduction

The financial crisis that engulfed the world at the close of 2008 has had dramatic implications for world oil and other energy supplies. For one, it has pushed oil prices to below US$50 a barrel from a record high of about US$150 per barrel. All of this took place within a matter of weeks. If the low price regime holds, it will have substantial implications for the urgent search for alternative and renewable energy. A clear tendency will be for crude oil suppliers to pump more oil in order to shore up the flow of their revenue. The Organization of the Petroleum Exporting Countries (OPEC) may seek to control the volume of oil in the market for similar reasons and hope that such curtailment would jack up prices. Will the oil corporations be uninterested bystanders in all this? Unlikely.

The oil sector has been a precursor of many conflicts since its ascendancy, and the correlation between incidence of wars and a rise in the cost of oil and attendant profits to the oil majors has already been established. With regard to present-day Africa, we can extend this argument to say that virtually all conflicts are about who has the right to exploit what resources are available. A member of the International Crisis Group had this to say about the fighting in the Democratic Republic of Congo, acknowledging the economic roots of the fighting: 'This is the absolute core of the conflict. What we are seeing now are beneficiaries of the illegal war economy fighting to maintain their right to exploit' (Swagler, 2008).

It is quite interesting to see that the usual laws of economics (the supply-and-demand theories) do not always work with the oil and gas industry. What we mean is that oil prices do not necessarily go up because there is a shortage of oil; and the prices do not necessarily fall because there is a glut in the market. This situation says something about the place of oil corporations' influence in today's governance systems, as well as about our fossil fuel-driven society. This theme runs through the discussion in this chapter.

There are, however, closely related factors that undeniably affect the profitability of the oil corporations: conflict and the threat of conflict. An analysis of the fact would more or less lead one to conclude that the oil and gas sector has been oiling conflicts throughout its stormy history. The arms-industrial complex cements the relationship between oil and conflicts. There are extensive arguments that show why this is so. These range from the fact that conflicts indicate possible disruptions of supplies, thus pushing up prices, to the argument that the arms industry directly thrives on selling arms and these sales are not at optimum levels in peacetime.

Oil revenue-dependent states become inexorably drawn into the trap of violence by the sheer need to maintain a firm grip on the levers of power. This happens because such states, although known as oil-producing states, are actually oil rent-collecting states. The oil corporations do all the production and the governments merely collect rents for the concessions. Such states have a regular inflow of petrodollars and enjoy windfalls when oil prices climb. As the petrodollars keep flowing in, these governments do not bother to rigorously collect tax from their citizens and mostly only those in formal institutional employment pay taxes – because they cannot avoid it. Since the vast majority of people do not directly contribute to the financial support of the structures of governance, there is a rather tenuous thread by which the governed are spurred to insist on accountability on the part of their governments. In order for government to maintain its grip on the people and power, when discontent creeps in, the tendency is for those in power to take the authoritarian route.

This brings in the arms industry, whose stock-in-trade is violence. It is reported that from 1984 to 1994, the military expenditures of OPEC countries as a percentage of their total budgets were three times that of developed countries, and up to ten times that of developing countries that were not members of OPEC (Ghazvinian, 2007, p106).

The question that our topic raises is: how local are the conflicts generated by this sector? If the conflicts are local, are the ripples circumscribed or do they spread regionally or even internationally? In this chapter, I will use Nigeria's Niger Delta as the localized arena. The Niger Delta has become an unmistakeable emblem of oil-sector-induced conflicts. The continued incidence of violence visited on peaceful communities here led to the expulsion of Shell Petroleum Development Company from Ogoni in 1993, for instance. In recent years, there has been a rise in armed groups fighting for space within the oil region. These fights have manifested themselves in the kidnapping of oil-sector workers, demand-and-receipt of ransoms and the destruction of oil pipelines. The upswing in violent confrontation contributed to the upward swing in crude oil prices globally, until somewhat checkmated by the financial crash that unravelled from the US in 2008.

The Niger Delta has been a historically neglected region of Nigeria. This neglect and the inability of the Nigerian state to deliver development and a stable democratic order have added deeper dimensions to the struggle for spoils in the region and the nation at large (Okonta, 2008, p181). The Niger Delta has become a parable from which new oil countries, especially of the East African Rift Valley area, are to draw lessons. If the states fail to draw the obvious lessons, it may be difficult for the region to stop drifting into more agonizing situations.

The rise of armed militias in Nigeria's oil fields has been traced to the setting up of gangs of thugs by politicians with the aim of fighting political opponents during election campaigns (Ghazvinian, 2007, pp54–55). With huge caches of arms in circulation and without tasks from their 'godfathers', the thugs turned their focus on illegal bunkering and other illegal ways of tapping into crude oil resources. It has been said 'that it would be no exaggeration to say that contemporary Nigerian politics are dominated by efforts to appropriate oil money' (Turner, 1980, p211). But there are some groups who claim loftier political ideals and agenda.

Even before the recent rise of armed gangs in the Niger Delta, popular resistance has often meant the intervention of the Nigerian military to dissipate such resistance. Over the years the Niger Delta has become the most militarized region of Nigeria. The role of the military has been suspect from as early as the 1990s as it regularly took an anti-people stance in the conflicts. It was responsible for the attacks on communities on behalf of oil corporations at different times, including Umuechem (1990), Ogoni (1993 to 1995), Odi (1999), Odioma (2005) and Ugborodo (2008). The oil companies, such as Shell, continue to provide 'logistical support' to government security forces in the Niger Delta, ostensibly to help secure their facilities. They also claim to be 'providing training to help them avoid human rights violations' (Shell, 2007, p24).

Apart from turning their guns against helpless local people, it has been revealed by none less than the Nigerian Chief of Defence Staff (CDS) Air Marshal Paul Dike 'that secret military information meant for the joint task force (JTF) fighting militancy in the Niger Delta region is leaked before it gets to the soldiers. The CDS also has information that some soldiers attached to the JTF encourage illegal bunkering in the area. He has also been informed of extortions and harassment of the people in the area by the soldiers' (Amaize et al, 2008).

In other situations, communities such as the Iwerekhan[1] have taken the path of litigation (against Shell's gas flares) and obtained unenforced victory, while others such as the Aja-Omaetan community in Warri North of Delta state chose to petition their governor for the cessation of gas flaring in their area. The people of Aja-Omaetan complain of intense heat from the gas flare point in the Dibi Oilfield, which is a mere 1km from the community. Chevron has been accused of deploying heavily armed security men to the area following the complaints of the people. In their petition to the Delta state governor, they stated that 'the effect of Chevron Nigeria limited gas flaring in

Dibi Field has taken a dangerous dimension as their farms have been burned beyond redemption, their land, air and surrounding water are overheated with its attendant associated unknown ailments now prevalent in the community'. Accordingly, the community urges and requested urgent intervention to 'stave off imminent bloodshed in the area' (Arubi, 2008).

The violent bent of the industry continues. In late November 2008, peaceful community members protesting Chevron's unwillingness to honour the terms of a memorandum of understanding that Chevron had entered into were shot at. In that incident, some Ugborodo community people, including at least one child, received injuries from gunshots (ERA, 2008a).

Politics of oil conflicts

Globalization has smiled broadly on the oil sector. It sits on top of the pile as a colossus, keeping a close eye on the turn of events, or often turning the events themselves. The power of the industry has been such that wars have become an opportunity, as seen in the case of Iraq. This is not really a new phenomenon, although the growth of the industry, including its mergers, has been phenomenal. According to Oilwatch International, since the wealth of oil corporations began to spread beyond the national markets of the countries in which they operated, they began to checkmate nation states throughout the world. This has resulted in wars between the big corporations and nation states. One such war began in the US when the first Standard Oil abridged the 'democratic' functioning of the state at the end of the 19th century. The confrontation spread to England, where the state struggled to control the powers of Royal Dutch Shell.[2]

The current situation in the Middle East, particularly Iraq, has shown in stark colours the interwoven nature of politics, conflicts and oil resource appropriation. It is a common belief that the entire Iraq war effort is all about oil, even though the trumpet had sounded the alarm over purported weapons of mass destruction. Those weapons turned out not to be hidden in some evil dictator's laboratory, but in the turrets of high-tech weapons manufactured by what has been termed the weapon dollar industry. This, however, may not be the entire question.

The wedlock between the industry and keepers of ill-gotten wealth is so rampant as to deserve no special treatment here. It is enough to say that states such as the US and Switzerland have become strategic recipients of slush funds from corrupt politicians and other crooks. The Iraqi situation has also shown how companies involved in the production or use of arms, as well as those engaged in the oil industry, benefit directly from systems that completely suspend even minimum concerns about basic standards of competitive bidding or any recourse to transparency. What this shows is that oil and power corrupt absolutely.

It should be instructive that at a time when oil fields have become hotbeds of conflicts and insurgency, it is precisely then that oil companies make record-breaking profits. Those involved in weapons trade, deconstruction/

reconstruction, private soldiers' contracting and the like also enjoy the boom. In the month of October 2006, when Iraqi civilian casualties of 3709 were recorded, a market analyst stated that Halliburton's quarterly profit was 'better than expected'. By the last quarter of 2006, it had enjoyed an inflow of up to US$20 billion from the Iraqi war alone. There is no need to keep track of its winning streak.

The question still remains: are energy conflicts ever local? It does appear that the answer to this question is negative. It has to be because the impact of the use of fossil fuels as the main energy source has global repercussions and effects. A glaring example is climate change. When man's contribution to the phenomenon is considered, the direct culprit can be traced home to an oil or gas well. The gasoline, diesel or gas may originate from Nigeria, Ecuador, Myanmar or Iraq; but the impact knows no national boundaries. The conflicts may be localized; but the impacts reach a wider world through a rise in energy costs, as well as through the displacement of local peoples and the ensuing drastic increase in environmental refugees.

The conflicts generated by the oil sector are sometimes played out of sight in some remote oil field, away from the media glare, and are thus allowed to go unquestioned. An example is the socio-economic conflicts generated when oil workers with 'large' sums of money in their pockets enter poor communities. In some cases, such as in Nigeria, oil companies speak of paying some community people monthly allowances even though the people do not work for them. This builds a body of company loyalists who work to thwart legitimate demands or expectations from individuals outside the sphere of this favour or even from the community. In the ensuing conflicts, people who ought to stay in solidarity are pulled apart and become antagonists – and the winner is the oil company. When these seething conflicts escalate, the way is open for both the oil corporations and the state to militarize the area. A militarized zone provides cover for the oil-sector companies to avoid accountability over oil spills and other acts of aggression against people.

A report on the 1973 Egypt–Israeli conflict and the expected impact upon oil supply gives an important message about the convergence of oil, politics and conflicts. The Saudis warned Mobil, Chevron and Texaco of an imminent attack on Israel by Egypt and the possibility of the use of oil as a political weapon in the conflict. The companies took steps to offset the effect on oil supplies by raising oil production in the first quarter of the year and this fully compensated for the eventual drop in the last quarter of the year (Shaxson, 2007, pp83–102).

Africa on the auction block

Oil raises hopes and dashes them all across the African continent. From Nigeria to Angola and Sudan, to many other countries, the story is largely the same. Oil and natural gas extraction cause environmental damage. Violent jockeying for the spoils continues to fuel instability, popular resentment and combustible anger (Donnelly, 2005).

As we have seen, oil and conflicts appear to be twins in today's world and the crisis must be viewed both in economic and political terms as a major for-profit venture. Understanding it through this filter is crucial to our seeing why we appear trapped in intractable murky waters and it may also help us to construct bridges over which we may come out of the malaise.

The lure of cheap oil has kept the rigs drilling through Africa, destabilizing governments and dividing communities. The poor continue to subsidize the costs of crude oil by the losses they suffer in environmental services, quality of life and extreme environmental degradation. This results in continued conflicts as opportunistic groups as well as gangs find space to extract financial gains from the system. As long as the profits flow, powerful forces in the world will delay serious efforts at developing readily available and cheap alternative energy.

It is doubtful if oil just happens to be found in places where there is conflict. Consider, for instance, the Darfur region of Sudan. While northern transnational corporations are often the culprit in terms of environmental despoliation, in Sudan a major culprit is Sinopec, the Chinese state oil corporation. In 1999, just as the first barrels of crude oil were shipped from Sudan, the war escalated between government forces and those of the Sudanese People's Liberation Army.

Today, oil has been found in the highly biodiverse lands of the Rift Valley in East Africa. Hopes and pulses are rising again. Oil majors are entering into contract arrangements and getting set to drill, even in ecologically sensitive areas. The African continent appears to have been divided among the oil corporations into concessions and plots for their exploitation and destruction. The scramble for Africa is seeing oil activities rapidly spreading in Eastern and Southern Africa. While oil and gas fever grips the tycoons in countries such as Tanzania, Mozambique, Madagascar, Chad, Mauritania, Ethiopia, Eritrea and Somalia, local communities are never consulted about what is about to hit them.

When the wells run dry

The world has finally woken up to the reality of climate change. Harsh weather events have shown that not even the most developed countries of this world can fully withstand these freak weather events. We are increasingly seeing floods, hurricanes, tornadoes, drought and wildfires of huge proportions. With these events, citizens around the world are experiencing increased incidence of disease, poverty, losses and untold hardships. Due to the fact that the climate crisis also manifests in restricted access to resources, incidents of conflicts are on the increase too. Africa is the worst hit and the most vulnerable.

It is generally believed that the world will soon witness a peak in oil production – the point at which the world exhausts half of all proven reserves.[3] Some experts estimate that Nigeria reached her own peak oil level two years ago.

However, the Nigerian government is planning to increase production from the current reported 2.5 million barrels a day to 5.2 million barrels a day by the year 2030. This increase is to meet the oil demands of the US as access to Middle East crude oil may get more difficult. It was recently reported that by early 2007, Nigeria became the third largest supplier of crude oil to the US, after Canada and Mexico (Akande, 2007). To underscore the strategic importance, the US is setting up a military command (AFRICOM) in the Gulf of Guinea, possibly to ensure that nothing disturbs the flow of oil from the region.

The Nigerian dream can only be realized if the geologic findings are wrong, as estimates already say that if production continues at the current rate, her oil will run out in less than 30 years. Escalating conflicts have resulted in the huge centralization of people or shut-ins through the rise of armed groups in the area. The power of oil to generate conflicts is growing in the continent. For more than 20 years, Cameroon and Nigeria have flexed muscles over ownership of Bakassi, an oil-rich marshland at the south-east tip of Nigeria. A ruling of the International Court of Justice that the land belongs to Cameroon does not appear to have resolved the crisis in those backwaters. Although the Nigerian flag has formally been lowered over the territory, tension remains in the creeks.

The almost desperate move by nations to find crude oil is leading to audacious activities, as well as innovations in the extractive fields. During recent years, the media reported Russia signifying ownership of the North Pole by planting a flag in the seabed there, 4.2km below the surface (BBC, 2007)! Claiming ownership by flag planting is an emblem of gaining victory through warfare. We may soon expect to hear that since the US planted a flag on the moon in 1969, they are the owners of that celestial body.

The troubling matter here is that while the countries of the North are securing energy resources, African countries are seeking to increase oil-related revenues. The fact that few resources are being spent on energy security efforts points to an ominous future. It is a scary thought to picture Nigeria as an energy-importing nation in the near future, with no prior arrangements on how to manage that scenario. This challenge is real for many African countries. One of the best cases of forward thinking can be seen in President Wade of Senegal's orchestration of the so-called *green* OPEC – a body of non-oil-producing African countries inaugurated in 2006. The green OPEC is made up of countries without crude oil, but who are poised to become exporters of agro-fuels possibly by converting cultivatable lands into fuel crop farms.

President Wade convened the first meeting of energy ministers from 13 nations to form the Pan-African Non-Petroleum Producers Association (PANPP), with the intention that it serve as a green version of OPEC. The members of PANPP aspire to become leaders in the field of biofuels and alternative energy strategies, following in Brazil's footsteps. But the development of a biofuels industry, particularly cellulosic biofuels made from agricultural wastes and prairie grasses (which President Bush touted in his State of the Union address) could take a decade or more to come to fruition. Africa, he stated, needs help today (Wade, 2006, pA15).

The green OPEC may not recognize it, but there is a major scramble for land in Africa today by agro-fuels promoters. This is already raising conflicts between communities and governments and promises to exacerbate food deficits and displacements in the region. This is happening because both land and the manpower that ought to go into food production is being diverted into the production of agro-fuels, not to meet local energy needs but for export. And there is simply not enough land in the tropics on which to cultivate enough agro-fuels to feed the tanks of the North.

The economics of war

However one measures it, oil is the most traded commodity in the 20th century. In no other period in man's history has the rate of consumption been so high. Petroleum products are so versatile that it is difficult to imagine life without such products as plastics made from this resource, and, for many, their continued use is taken for granted. In fact, from 1983, plastic and related materials have been produced at higher volumes than steel. Today, a huge variety of domestic products, ranging from dyes to nail polish, are derived from petroleum.

An additional reason for the profitability of the sector is that few other commodities have generated so much surplus value. I contend that this surplus value is a result of the industry operating with impunity, paying scant attention to the environment and to the people who live in their areas of operation. Thus, the energy produced by the sector remains heavily subsidized by faceless, voiceless millions whose waters, lands and livelihoods are circumscribed on a daily basis.

The path of crude oil development has been strewn with skeletons and soaked in human blood across the world. The ongoing case in Nigeria is a glaring example, and the case of Angola is still fresh in memory. Naomi Klein has expertly exposed the issue of the profitability of disasters in her new book. She states that 'With resource scarcity and climate change providing a steadily increasing flow of new disasters, responding to emergencies is simply too hot an emerging market to be left to the nonprofits – why should UNICEF rebuild schools when it can be done by Bechtel, one of the largest engineering firms in the US?' She also asks the question: 'Why deploy UN peacekeepers to Darfur when private security companies like Blackwater are looking for new clients?' (Klein, 2007, p15).

Diving into the deep blue sea

While both the government and oil companies are the beneficiaries of the crisis raging in the oil fields, enjoying huge profits and windfalls, both are equally vulnerable to the challenge of access to oilfields. One Nigerian activist (in private conversation) posited that the Nigerian government is a victim of disaster capitalism. The new government is caught in the web of supremacist gangs, engaged in the kidnappings of oil workers and the abductions of children and parents of politicians. Unless they take steps to look away from short-term profits and to work for the security of the environment, livelihoods and the rights of the people to live in a way that is favourable to their development (according to the 2008 African Charter on Human and Peoples' Rights, Article 24), oil companies and governments are unlikely to continue to enjoy ongoing oil revenues.

Offshore activities are being intensified in the Gulf of Guinea: Nigeria, Sao Tomé and Principe, Equatorial Guinea, Angola, and newcomers such as Ghana and Sierra Leone. But even access to offshore installations is becoming less secure. Deep sea and remote offshore locations pose financial risks besides the physical ones.

One factor that makes offshore drilling and platforms so attractive is their perceived security and the lack of accountability to 'host communities' since no one can easily monitor their activities out at sea. However, a recent attack of one of the Bonga floating production storage and offloading (FPSO) facilities, 75km offshore, allegedly by Niger Delta militants, has made the claim of security a mere dream (Arubi et al, 2008). Perhaps for the oil companies, ultimate security will only come with the landing of US troops in one of the nations under their AFRICOM, which came into full operation as a unified command in October 2008.

Following the attack on the World Trade Centre on 11 September 2001, US policy towards Africa has been largely built on oil and terrorism (Feller, 2003, p149). The US is seen as building deeper military ties with neoliberal protégés such as Nigeria, South Africa and Mozambique, working to weaken OPEC by trying to persuade her allies to leave the cartel, and doing this, as an industry operative put it, by peeling off certain countries. By 2002, President Bush's Africa policy was already characterized as 'build the military and extract the oil' (Montague, 2002)

About 42 per cent of Nigeria's crude is exported to the US. Together with the prodigious fields of the rest of the Gulf of Guinea, about 25 per cent of the crude needs of the US are to be met from Nigeria by 2015. Already the US has suggested that the Gulf of Guinea is an ungoverned space, implying that she may be intending to provide this governance through military bases. It was initially believed that this would come through AFRICOM; but it now appears as if the troops will be engaged in community development services. Such services will include building schools and digging wells. However, many observers believe that, ultimately, the command seeks to protect oil resources to ensure supplies to the US and to counter Chinese incursions into the region.[4]

In a policy brief, the Trans-Africa Forum described AFRICOM as 'the newest iteration of a failed foreign policy agenda driven by exploitation of natural resources and the expansion of the Global War on Terror' (TransAfrica Policy Brief, 2008). The body describes the US foreign policy in Africa as 'one of unilateral prioritizing of US government and corporate interests over African development'.

Beaten by all sides

Fence-line communities suffer great impacts from energy production activities. This is very stark in communities where oil is extracted in the Niger Delta. Natural gas associated with crude oil extraction is burned in inefficient and highly injurious infernos on a continuous basis. This routine flaring has been declared an illegal activity by a high court in Nigeria and the Nigerian Senate is making efforts to formally outlaw it, probably because both the oil corporations and the government have ignored the court order. This flaring wastes US$15 million worth of gas daily, pumping massive quantities of greenhouse gases as well as toxic substances into the air. There is no breath of fresh air near these flares. They cause asthma, bronchitis, cancers and blood disorders. They also pour acid rain on the land, vegetation, buildings and the people (Conant and Fadem, 2008). These communities suffer all of these impacts and more, but have no electricity and no benefits whatsoever from the oil and gas being produced and/or wasted in their communities.

Over 100 flare sites in the Niger Delta belch 400 million tonnes of carbon dioxide equivalent into the atmosphere annually. Through flaring, Nigeria lost an estimated US$72 billion in government revenues for the period of 1970 to 2006, or US$2.5 billion annually (ERA, 2008b).

In South Africa, communities living near refineries are severely affected by the toxic fumes and chemicals released from them. In some of the communities, especially the ones in South Durban, it is difficult to find a family without at least one person having a severe respiratory disease.[5] Conflicts with local communities here are a daily reality as people struggle for an illusive breath of fresh air. These conflicts are not limited to locations in Africa. They are replicated in Ireland over Shell's pipelines, in Myanmar over Total's pipeline and in refinery fence-line communities in the US.

One other way for oil companies to attempt to douse energy conflicts is by ensuring local ownership of (or buying off) the local community. However, the provision of school blocks and clinics has not been effective in resolving conflicts. One community activist made a poignant remark when he said that he would rather stay healthy than endure pollution and stay sick in a well-equipped hospital.

Communities affected by oil extraction in Africa, Asia, Latin America and elsewhere across the world are asking questions. A web of resistance is building up. A movement is growing demanding that crude oil be left in the ground as a technology-free, direct and effective means of carbon sequestration and as a real contribution towards fighting climate chaos. They want to know why oil should still be drilled in their land when the process and the resource are poisoning their lands, waters and atmosphere. They want to know why they must live with the oil spills, gas flares and the moral, economic and social dislocations that trail the industry. They want to know why they must be silent in the face of serious deforestation caused by the oil industry and in the face of toxic chemicals from refineries. They are asking that the crude oil be left in the ground where it rightly belongs.

Beating a just path forward

In a discourse such as this it may be tempting to throw up one's hands in despair that it is too late and there is nothing that one can do to turn the tide. This must not be the case. Actions at the grassroots level have sought ways to beat a path to the sustainable use of resources in the Niger Delta and elsewhere. The major plank used by groups such as Environmental Rights Action (ERA) – the Nigerian chapter of Friends of the Earth International – has been that of community education. ERA believes that creation of wealth without knowledge and popular participation cannot lead to sustainable development.[6]

ERA has pursued this through efforts to democratize development while protecting the environment from wanton destruction. Practically, this is done through community education on environmental and human rights monitoring,[7] as well as training on the use of legal tools for securing rights and taking steps to defend their environment. Environmental education includes environmental health information and training. ERA uses resources such as *A Community Guide to Environmental Health* by the Hesperian Foundation and the *Community Environmental Monitoring Handbook*[8] by the Akwa Ibom Research and Information Organization. Environmental monitoring is a key tool that helps to keep harmful activities and events in the public view, thus forcing responses from otherwise impervious polluters.

A clean-up of the highly polluted oil and gas fields will help to assuage hurt nerves and also make it possible for local people to regain their means of livelihoods – tightly linked to available environmental resources. A 1997 report by the World Wide Fund for Nature (WWF) estimated that cleaning up the Niger Delta environment would cost up to UK£4 billion (Okonta and Oronto, 2001, p272). The oil corporations are estimated to have pumped out oil worth over US$350 billion since operations started in the region. It is our considered estimate that this total sum will be needed just to begin an auditing of the degradation and commencement of efforts to restore the wrecked physical environment of the Niger Delta, without even considering damage to the political, spiritual and social spheres. But it would be a good start. And time is running out.

Notes

1 The community obtained a landmark victory at the Federal High Court sitting in Benin City, Nigeria, on 14 November 2005 with a declaration that gas flaring is a violation of the human rights of the local people, as well as being a monumental waste of resources. The court more or less declared gas flaring an illegal activity.

2 *Multinational Monitor* (2007) 'Editorial: The end of oil', *Multinational Monitor*, Washington, January/February, p6. This issue of the *Multinational Monitor* illustrates, among others things, that the 'Corporate control of energy policy and energy resources, especially in the United States, the country that consumes more energy than any other, is the single greatest obstacle to slow and hopefully reverse the world's headlong rush to disaster.'

3 Oilwatch International, *World Oil Atlas*, Quito.

4 See www.resistafricom.org for more information on this and related topics.
5 Visit GroundWork/Friends of the Earth South Africa website for more details: www.groundwork.org.za.
6 See more on ERA's philosophy at www.eraction.org.
7 ERA works in collaboration with other Nigerian and African NGOs and networks. Such networks include the Green Alliance Nigeria – a network of environmental NGOs in Nigeria – and the Host Communities Network (HoCoN). HoCoN is a network of communities affected by the extractive industry activities. The HoCoN concept blunts the strategy of government and transnational corporations to divide communities hosting their activities from those who merely suffer the impacts.
8 See www.hesperian.org.

References

Akande, L. (2007) 'Nigeria is third largest oil exporter to US', *The Guardian*, Lagos, 13 June 2007, www.guardiannewsngr.com/news/article02

Amaize, E., K. Omonubi and U. Kalu (2008) 'Comments on Africa Charter on Human and Peoples' Rights, Article 24', *Vanguard*, www.vanguardngr.com/index.php?option=com_content&task=view&id=11187&Itemid=0

Arubi, E. (2008) 'Community accuses Chevron of intimidation', *Vanguard*, www.vanguardngr.com/index.php?option=com_content&task=view&id=13241&Itemid=44

Arubi, E., O. Kingsley Abuja and U. Kalu (2008) '"We have reports of soldiers encouraging militancy", says CDS, Dike', *Vanguard*, Lagos, 27 September 2008, www.vanguardngr.com/content/view/17941/41/

BBC (2007) 'Russia plants flag under North Pole', news.bbc.co.uk/2/hi/europe/6927395.stm

Conant, J. and P. Fadem (2008) *A Community Guide to Environmental Health*, Hesperian Foundation, New Earth Rising Berkeley, CA

Donnelly, J. (2005) 'Burdens of oil weigh on Nigerians Ecological harm, corruption hit hard', *Boston Globe*, 3 October 2005, www.boston.com/news/world/africa/articles/2005/10/03/burdens_of_oil_weigh_on_nigerians/

ERA (Environmental Rights Action) (2008a) *Chevron Security Shoots Protesters at Aruntan – Ugborodo Community*, Field Report 189, Benin, 25 November 2008, www.eraction.org/index.php?option=com_content&view=article&id=141:field-report-189-chevron-security-shoots-peaceful-protesters-at-aruntan-ugborodo-community-&catid=5

ERA (Environmental Rights Action) (2008b) *Gas Flare Factsheet*, December, ERA, Nigeria

Feller, J. (ed) (2003) *Power Trip – US Unilateralism and Global Strategy After September 11*, Seven Stories Press, New York, NY, p149

Ghazvinian, J. (2007) *Untapped: The Scramble for Africa's Oil*, Harvest Inc, Fort Washington, PA, pp54–55, 106

Klein, N. (2007) *The Shock Doctrine*, Penguin Books, London, p15

Montague, D. (2002) *Africa: The New Oil and Military Frontier*, Arms Trade Resource Center Update, 20 September, www.resistafricom.org

Multinational Monitor (2007) 'Editorial: The end of oil', *Multinational Monitor*, Washington, January/February, p6

Okonta, I. (2008) *When Citizens Revolt: Nigerian Elites, Big Oil, and the Ogoni Struggle for Self-Determination*, Ofirima Publishing House Ltd, Port Harcourt, p181

Okonta, I. and D. Oronto (2001) *Where Vultures Feast*, Environmental Rights Action, Benin, p272
Shaxson, N. (2007) *Poisoned Wells: The Dirty Politics of African Oil*, Palgrave, New York, NY, pp83–102
Shell (2007) *Responsible Energy: The Shell Sustainability Report 2007*, www.sustainabilityreport.shell.com/2007/servicepages/welcome.html, p24
Swagler, M. (2008) 'Behind the war in the Congo', *Socialist Worker*, November 2008, www.socialistworker.org/2008/11/25/behind-war-in-the-congo
TransAfrica Policy Brief, AFRICOM (2008) *The 'Freedom Agenda' on the African Continent*, June 2008, www.transafricaforum.org
Turner, T. (1980) 'Nigeria: Imperialism, oil technology and the Comprodor state', in P. Nore and T. Turner (eds) *Oil and Class Struggle*, Zed Books, London, p211
Wade, A. (2006) 'Africa over a barrel', *Washington Post*, Saturday, 28 October

Part II
Climate Insecurity: A Challenge to Peace, Security and Development

7
Creating a Climate of Security: The Latest Science and Acceptable Risk

Andrew Higham[1]

Introduction

The climate system is being transformed by human civilization into one that poses much greater risks, with potentially catastrophic consequences for society and ecosystems. The world is warming faster than expected and significant impacts are appearing on all continents. During 2007 and 2008, polar ice thickness and extent reached record low levels (reducing, on average, by 9.1 per cent per year from 1979 to 2006). Mountain glaciers and snow cover have declined in both hemispheres and widespread decreases in glaciers and ice caps have contributed to sea-level rise. In the Southern Hemisphere, climatic extremes are being exacerbated with longer and more profound droughts, more pronounced flood events and fire risks.

Greenhouse gas emissions continue to increase at an ever increasing pace – concentrations of greenhouse gases in the atmosphere have increased faster during the last ten years than at any time since continuous measurements began in 1960 (NASA, 2008).

If trends continue, the Earth's average temperature could increase by a further 4°C to 6°C by the end of this century (Hare, 2009), and based upon the paleo-data, there is a risk of precipitating devastating climatic shifts where the climate system 'flips' into a much different, less habitable state (Lenton et al, 2008). Even if we are moderately successful in containing emissions growth and, hence, global temperature increases to within 3°C, there is still a significant chance of massive human health and welfare impacts, as well as major disruptions to ecosystems. Knowledge about these impacts can aid in efforts to adapt to climate change; but avoidance should always remain the first line of defence in climate policy.

The Fourth Assessment Report of the Intergovernmental Panel on Climate Change (IPCC, 2007b) shows that reducing greenhouse gas emissions by 25 to 40 per cent below 1990 levels by 2020, and by 80 per cent by 2050 would give us a 50 per cent chance that the global temperature increase could be kept to within 2°C. However, reducing emissions by 80 per cent by 2050 will not eliminate some serious risks and damages.

Other chapters in this book highlight the consequential security challenges that are associated with the projected impacts of climate change. In this chapter we examine the latest Earth system science to characterize the major risks to society posed by global warming.

Projected global average temperatures

Average global atmospheric warming over the last century is almost 0.8°C. According to the National Aeronautics and Space Administration (NASA), the global average surface temperature has risen by approximately 0.6°C in just 30 years (NASA, 2008). Twelve of the warmest years on record have all occurred in the last 13 years. Every year from 2001 to 2007 was warmer than every year from 1991 to 2000, with the exception of 1998 (Hadley Centre, 2008).

During the 21st century, the IPCC projected that global average temperature could increase within the range of 1.1°C to 6.4°C above 1990 levels, which is about 1.6°C to 6.9°C above global average temperatures prior to 1750 (IPCC, 2007b).

Temperature increases are projected to vary significantly at different latitudes, with much greater average increases in temperature in the higher latitudes. Higher altitudes and areas further inland will also experience greater temperature increases, and the frequency and severity of extreme temperatures is also likely to be very significant.

Observed temperature increase is in the upper part of the range projected by the IPCC and emissions growth is currently greater than the most emissions-intensive scenario of the IPCC. Unless the trends of the past 30 years are an aberration, and if a global solution remains illusive, the worst case climate change scenario is unfolding.

Based on climate model studies and advanced studies of climatic and Earth system history, Hansen et al (2008) conclude that additional global warming of about 1°C is likely to be dangerous. Makiko Sato of Columbia's Earth Institute concludes that 'the temperature limit implies that carbon dioxide exceeding 450 parts per million is almost surely dangerous'. As James Hansen states (Hansen et al, 2008):

> *If global emissions of carbon dioxide continue to rise at the rate of the past decade, this research shows that there will be disastrous effects, including increasingly rapid sea-level rise, increased frequency of droughts and floods, and increased stress on wildlife and plants due to rapidly shifting climate zones.*

Projected sea-level rise

Robust estimates of projected sea-level rise are not available. The IPCC (2007b) could not estimate all of the contributions of global warming in projecting sea-level rise, mostly due to uncertainties with the response of the Greenland and Antarctic ice sheets. Given recent evidence that the stability of Greenland and West Antarctic ice sheets are weakening and eroding faster than previously expected, it is likely that the IPCC estimate of 0.18m to 0.59m by 2100 is a significant underestimate.

The most recent data indicate that average sea-level rise is approximately 3.4mm per year, compared to 3.1mm per year between 1993 to 2003, and 1.8mm per year from 1961 to 2003 (Church et al, 2008; Rahmstorf, 2007). Analysis shows that the rate of sea-level rise in the 20th century could be the highest for the past 5000 years, increasing at almost twice as much since 1993 than the average in the past century (ACE CRC, 2008). Like temperature, trends in sea-level rise are at the high end of the IPCC projections (White et al, 2008). Estimates that better reflect potential ice sheet flows are now showing larger projections of sea-level rise. For example, Rahmstorf (2007) projects sea-level rise of between 0.5m to 1.4m above the 1990 level by 2100, which is more than double the IPCC estimate.

Of course, sea-level rise will not be uniform. In some locations, sea levels will increase greater than in others. It is, however, very difficult to project sea-level rises at a regional or local scale. The climate-change-driven increases in the height or frequency of extreme sea-level events, such as very high tides and storm-related surges, are already threatening some coastal and small island settlements, with relocations of some communities occurring (Green, 2008) or being planned in the Pacific region (e.g. Torres Strait Islands, Vanuatu, Maldives, Kiribati and Tuvalu).

Precipitation

The frequency of heavy precipitation events has increased over most land areas. Significantly increased precipitation has been observed in eastern parts of North and South America, Northern Europe and Northern and Central Asia. There is also observational evidence for an increase of intense tropical cyclone activity in the North Atlantic since about 1970.

Drying has also been observed over large regions, including the Sahel, the Mediterranean, Southern Africa and parts of Southern Asia. In Africa's large water catchments of Niger, Lake Chad and Senegal, total available water has decreased by 40 to 60 per cent, and desertification has been worsened by lower average annual rainfall, runoff and soil moisture, especially in Southern, Northern and Western Africa.

Precipitation can be expected to continue to decrease in most subtropical land regions, but to increase in the high latitudes. The IPCC (2007a) found with 'high confidence that many semi-arid areas (e.g. Mediterranean Basin, western US, Southern Africa and north-east Brazil) will suffer a decrease in water resources due to climate change'. By the 2050s, it is projected that there will be less annual river runoff and water availability in dry regions in the mid latitudes and tropics, but an increase in high latitude regions and in some tropical wet areas.

Climatic extremes

Weather extremes are already having significant impacts upon human and natural systems. Human activity has led to the increased temperature. It is likely that heat waves are more frequent – the chances of a repeat of the devastating 2003 European heat wave event has quadrupled due to climate change (Stott et al, 2004; Gutowski et al, 2008). Indeed, by the 2040s, more than half of European summers will be hotter than the summer of 2003, and by the end of this century, a summer as hot as that of 2003 will be considered 'mild' (Stott et al, 2004).

Increases in heavy rainfall events have been observed since 1950, with some increase in flooding irrespective of increases or decreases in total precipitation. It is also likely that there has been an increase in very rare extreme precipitation events. According to Laurens Bouwer of the Dutch Institute for Environmental Studies, Jakarta faces a potential doubling of economic losses until 2015, and in Bangladesh the amount of people potentially affected by flooding will increase by 35 per cent over the same period (Bouwer et al, 2007).

By definition, extreme events are rare, local and may be short lived, and observational data are limited. Therefore, trends are difficult to detect. Projecting the future frequency and severity of extreme events is also one of the more difficult aspects of climate research. Extreme events are, however, crucially important for understanding the likely damages and costs of climate change. The costs tend to have a non-linear response to severe weather events – relatively small changes in the frequency and severity of weather events usually cause disproportionately larger damages. And as cyclone Katrina in the US, the 2003 European heat waves and Australia's recent massive bushfires and extensive droughts have demonstrated, even resilient and wealthy nations with the capacity to adapt and respond are highly vulnerable to increases in extreme events.

The IPCC (2007b) concluded that more frequent extreme heat events and heat waves, more intense and prolonged (although not more frequent) tropical cyclones, and heavier precipitation and flooding can be expected in many regions.

Tipping points

'Warming could lead to some impacts that are abrupt or irreversible, depending upon the rate and magnitude of the climate change' (IPCC, 2007a).

Unfortunately, history shows that the climate system does not always respond in a gradual, linear manner to increased concentrations of heat-trapping greenhouse gases. Rather, the system can respond slowly, resisting change and obscuring the underlying march towards a critical threshold.

Once the threshold is crossed, the system will respond relatively quickly, recalibrating into a new, very different state with massive implications for life (Pearce, 2007).

Furthermore, the emission patterns of today will 'lock in' high global average temperatures for the next 1000 years. Solomon et al (2009) show how:

> *... climate change that takes place due to increases in carbon dioxide concentration is largely irreversible for 1000 years after emissions stop. Following cessation of emissions, removal of atmospheric carbon dioxide decreases radiative forcing, but is largely compensated by slower loss of heat to the ocean, so that atmospheric temperatures do not drop significantly for at least 1000 years.*

There is now stronger evidence that the world's oceans and forests are absorbing less of the carbon dioxide released by human activity. Significant weakening of carbon absorption of potential Earth systems compounds increased emissions rates. There is also the potential for the reversal of some carbon absorption systems such that they become net sources of emissions. For example, should drought frequency in the Amazon increase significantly, it is possible that the largest terrestrial sink of carbon dioxide in the world will become a net source of carbon emissions (Phillips et al, 2009).

Lenton et al (2008) assessed 15 potential tipping elements in the Earth system and ranked the likelihood that each may arise in this century. They exclude those tipping elements unlikely to be triggered within the 21st century, leaving eight tipping elements ranked as either:

1 highly sensitive with the smallest uncertainty;
2 intermediate sensitivity with largest uncertainty; and
3 low sensitivity with intermediate uncertainty.

They also assess how much additional global warming would be necessary to trigger the tipping point. The results are summarized in Table 7.1.

Smith et al (2009) summarize the increasing risks associated with climate change by comparing known risks with those assessed in the IPCC's third assessment report of 2001. Figure 7.1 compares these risks using a visual display across five issues of concern originally used in the 2001 IPCC assessment. The transition from white through grey to black indicates increasing levels of risk, which is a function of both the potential severity and likelihood of impacts as global average temperatures increase.

Many have argued that it could be an option to accept some level of risk and warming, and deal with the consequent climate changes through measures that build resilience, allowing us to adapt to climate change. This view has become institutionalized within the negotiating process, with many of the analyses commissioned by the Conference of the Parties under the Bali Action

Plan estimating finance and technology needs according to a 500 to 550 part per million carbon dioxide equivalent stabilization (e.g. UNFCCC, 2007).

Metz et al (2007) give a 50 per cent chance that a 550 part per million carbon dioxide equivalent stabilization will cause a 20 and 30 per cent loss of all species on Earth with almost half of the world's population being at risk of water shortages, 0.25 billion people exposed to potential health problems, and hundreds of millions facing food shortages and coastal inundation.

Decrease in Antarctic sea ice and near complete disappearance of summer sea ice in the Arctic towards the end of this century will devastate ice-dependent ecosystems and extend into the interior of the bordering continental regions of Russia, Canada and Alaska, risking runaway methane emissions released from thawing permafrost, and the release of methane hydrates currently trapped deep in the ocean as temperature increases extend into the ocean.

In some of the temperate zone and in drier tropical areas, water availability will decrease and ecosystems will become stressed. Some ecosystems are already suffering and massive tree deaths are being recorded. In Western Australia, rainfall has declined by 20 per cent in 30 years resulting in a 50 per cent reduction in water availability with a further 20 per cent predicted (Government of Western Australia, 2008). Glaciers in regions such as central Asia and the Himalayan and Tibetan plateau are melting faster than expected

Table 7.1 *Tipping elements ranked according to likelihood*

Tipping element	*Additional global warming required (°C relative to 1990)*	*Transition timescale (years)*	*Ranking (1–3)*	*Potential impacts*
Arctic summer sea ice	0.5°C–2°C	~10	1	Possible complete loss of ecosystem and amplified warming; accelerated permafrost loss
Greenland ice sheet	1°C–2°C	>300	1	Possible 6m–7m sea-level rise
West Antarctic ice sheet	3°C–5°C	>300	2	Possible 4m–5m sea-level rise
Atlantic thermohaline circulation	3°C–5°C	~100	3	Large-scale climate changes globally
El Niño Southern Oscillation	3°C–6°C	~100	2	Extensive drought events
Indian summer monsoon	Not applicable	~1	Not ranked as not comparable	Either potential increased or decreased rainfall
Sahara/Sahel and West African monsoon	3°C–5°C	~10	2	Either potential increased or decreased rainfall
Amazon rainforest	3°C–4°C	~50	2	Massive extinction event, decreased rainfall and amplified warming
Boreal forest	3°C–5°C	~50	2	Loss of ecosystem

Source: adapted from Lenton et al (2008)

and, as discussed in Chapter 10, the long-term loss of these water resources could affect billions of people.

Impacts upon water regimes are already affecting agriculture and food supply, which will be exacerbated in some regions, mostly the least developed nations, even if warming can be kept within 2°C to 3°C.

Hansen et al (2008) call for stabilization at 350 parts per million of carbon dioxide, significantly below present levels of nearly 390 parts per million of carbon dioxide, which corresponds to limiting global average temperature increases to 1.7°C above pre-industrial levels.

According to Hare (2009), there is a '75 per cent risk that stabilizing greenhouse gas concentrations at 550 parts per million would lead to warming exceeding 2°C'. However, 'for a concentration pathway that peaks at 475 parts per million carbon dioxide equivalent and then drops to stabilize at 400 parts per million carbon dioxide equivalent, there would be about a 20 per cent chance of exceeding 2°C'.

He suggests that a plausible and risk-adverse emissions reduction pathway is to keep global average temperature increases below 2°C and also have a decent chance of holding temperature increases below 1°C. This requires fossil carbon dioxide emissions to approximate zero in 2050 and an 85 per cent reduction in all greenhouse gases from 1990 levels by 2050, peaking in 2020 and 'going negative' by 2075.

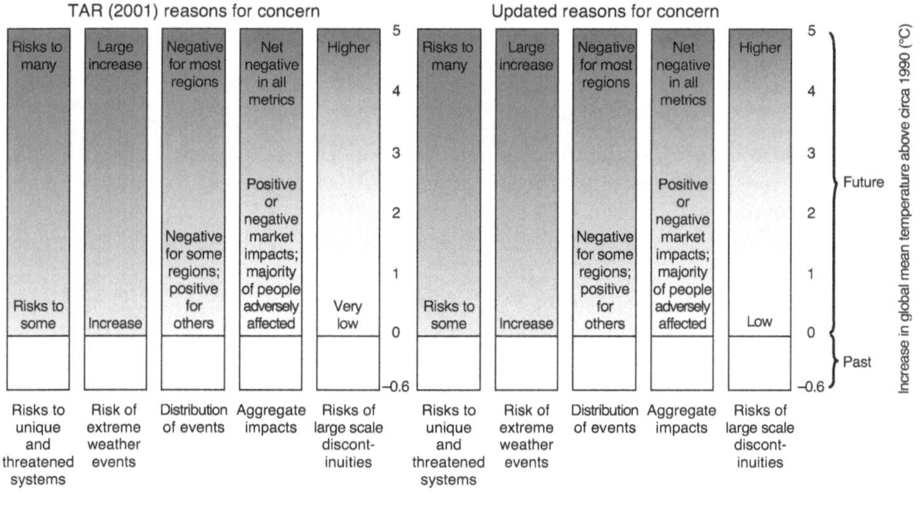

Source: Smith et al (2009)

Figure 7.1 *Risks from climate change: Comparison of estimates of risks for each reason of concern as assessed by the IPCC in 2001 and a revised assessment by Smith et al (2009)*

This last point is crucial. A risk-adverse strategy for climate change relies on creating a global economy that actively draws down the concentrations of greenhouse gases – that is, one that absorbs more than it emits. This can only be achieved through massive reforestation and new energy generation technologies such as biomass energy with carbon capture and storage and technologies that directly remove carbon dioxide (or other greenhouse gases) from the atmosphere. It suggests that the shared vision for the convention should be a carbon-neutral global economy soon after 2050.

The marginal costs of such a strategy are likely to be high; but relative to the residual costs and risks of a 2°C to 3°C outcome, it is, in our judgement, a cost worth bearing. Indeed, the cost is expected to be less than 1 to 2 per cent of gross domestic product (GDP) (see Chapters 9 and 24). Risk is a function of both the likelihood of an event occurring and the severity of the event. What is an acceptable level of risk is a matter for judgement – it is fundamentally a moral question. Should we accept the additional cost and maintain a safe climate, or should we shoot for 2°C to 3°C and accept the loss of much of the world's coral reefs, extinction of many species, displacement of millions of people, significant health risks, and loss of or disruption to some entire nations? Add to this the more recent evidence that global warming of 2°C to 3°C may not avoid catastrophic tipping points over the longer term and I am left with no doubt which is the most rational global strategy.

Note

1 The author wishes to thank Dr Bill Hare for the inspiration for this chapter.

References

ACE CRC (Antarctic Climate and Ecosystems Cooperative Research Centre) (2008) *Position Analysis: Climate Change, Sea-Level Rise and Extreme Events – Impacts and Adaptation Issues*, ACE CRC, Hobart, p5

Bouwer, L., R. Crompton, E. Faust, P. Hoppe and R. A. Pielke Jr (2007) 'Confronting disaster losses', *Science*, vol 318, p753

Church, J.A., N.J. White, T. Aarup, W.S. Wilson, P.L. Woodworth, C.M. Domingues, J.R. Hunter and K. Lambeck (2008) 'Understanding global sea levels: past, present and future, *Sustainability Science*, Special Feature. Original article, doi:10.1007/s11625-008-0042-4

Government of Western Australia (2008) *Making Decisions for the Future: Climate Change: The Premier's Climate Change Action Statement*, Perth, Western Australia

Green, D. (2008) *Impacts on the Health of Remote Northern Australian Indigenous Communities*, Report to the Garnaut Climate Change Review, www.garnautreview.org.au

Gutowski, W. J., G. C. Hegerl, G. J. Holland, T. R. Knutson, L. O. Mearns, R. J. Stouffer, P. J. Webster, M. F. Wehner and F. W. Zwiers (2008) 'Causes of observed changes in extremes and projections of future changes', in T. R. Karl, G. A. Meehl, C. D. Miller, S. J. Hassol, A. M. Waple and W. L. Murray (eds) *Weather and Climate Extremes in a Changing Climate: Regions of Focus – North America, Hawaii, Caribbean, and US Pacific Islands*, Synthesis and Assessment Product 3.3, US Climate Change Science Program, Washington, DC, pp81–116

Hadley Centre (2008) *Observations Dataset*, www.hadobs.metoffice.com

Hansen, J., M. Sato, P. Kharecha, D. Beerling, R. Berner, V. Masson-Delmotte, M. Pagani, M. Raymo, D. L. Royer and J. C. Sachos (2008) 'Target atmospheric carbon dioxide: Where should humanity aim?', *Journal of Atmospheric Science*, vol 2, pp 217–231

Hare, W. L. (2009) 'A safe landing for the climate', in World Resources Institute (ed) *State of the World 2009: Into a Warming World*, WRI, Washington, DC

IPCC (Intergovernmental Panel on Climate Change) (2001) *IPCC Third Assessment Report*, UN, Geneva

IPCC (Intergovernmental Panel on Climate Change) (2007a) 'Summary for policy makers', in *Climate Change 2007: The Physical Science Basis*, Cambridge University Press, Cambridge, UK

IPCC (2007b) *Climate Change 2007: Fourth Assessment Report of the Intergovernmental Panel on Climate Change*, Cambridge University Press, Cambridge, UK, pp173–210

Lenton, T. M., H. Held, E. Kriegler, J. W. Hall, W. Lucht, S. Rahmstorf and H. J. Schellnhuber (2008) 'Tipping elements in the Earth's climate system', *Proceedings of the National Academy of Science*, vol 105, no 6, pp1786–1793

Metz, B., O. Davidson, P. Bosch, R. Dave and L. Meyer (eds) (2007) *Climate Change 2007: Mitigation*, Contribution of Working Group III to the Fourth Assessment Report of the Intergovernmental Panel on Climate Change, Cambridge University Press, Cambridge and New York

NASA (National Aeronautics and Space Administration) (2008) *GISS Dataset*, data.giss.nasa.gov

Pearce, F. (2007) *With Speed and Violence: Why Scientists Fear Tipping Points in Climate Change*, Beacon Press, Boston, MA

Phillips, O. L., O. C. Aragao, S. L. Lewis, J. B. Fisher, J. Lloyd, G. Lopez-Gonzelez, Y. Maldi, A. Monteagudo, J. Peacock, C. A. Quesada, G. van der Heijden, S. Almeida, I. Amaral, L. Arroyo, G. Aymard, T. R. Baker, O. Banki, L. Blanc, D. Bonal, P. Brando, J. Chave, A. C. Alves de Oliveria, N. D. Cardozo, C. I. Czimczik, T. R. Feldpausch, M. A. Frietas, E. Gloor, N. Higuchi, E. Jimenez, G. Lloyd, P. Meir, C. Mendoza, A. Morel, D. A. Neill, D. Nepstad, S. Patino, M. C. Penuela, A. Prieto, F. Ramirez, M. Schwarz, J. Silva, M. Silveira, A. S. Thomas, H. ter Steege, J. Stropp, R. Vasquez, P. Zelazowski, A. D. Esteban, S. Andelman, A. Andrade, K.-J. Chao, T. Erwin, A. Di Fiore, E. Honorio, H. Keeling, T. J. Killeen, W. F. Laurence, A. P. Cruz, N. C. A. Pitman, P. N. Vargas, H. Ramirez-Angulo, A. Rudas, R. Salamao, N. Silva, J. Terborgh and A. Torres-Lezama (2009) 'Drought sensitivity of the Amazon rainforest', *Science*, 6 March, pp1344–1347

Rahmstorf, S. (2007) 'Response to comments on "A semi-empirical approach to projecting future sea level rise"', *Science*, vol 317, p1866d

Smith, J. B., S. H. Schneider, M. Oppenheimer, G. W. Yohe, W. L. Hare, M. D. Mastrandrea, A. Patwardhan, I. Burton, J. Corfee-Morlot, C. H. D. Magadza, H.-M. Füssel, B. Pittock, A. Rahman, A. Suarez and J.-P. van Ypersele (2009) 'Assessing dangerous climate change through an update of the Intergovernmental Panel on Climate Change (IPCC) "reasons for concern"', *Proceedings of the National Academy of Science*, www.pnas.org/cgi/doi/10.1073/pnas.0812355106

Solomon, S., G.-K. Plattner, R. Knutti and P. Friedlingstein (2009) 'Irreversible climate change due to carbon dioxide emissions', *Proceedings of the National Academy of Science*, vol 106, no 6, pp1704–1709

Stott, P. A., D. A. Stone and M. R. Allen (2004) 'Human contribution to the European heatwave of 2003', *Nature*, vol 432, no 7017, pp610–614

UNFCC (2007) *Investment and Financial Flows to Address Climate Change*, UNFCC, Bonn

White, N. J., S. Jevrejeva, S. J. Holgate, J. A. Church, W. R. Gehrels (2008) 'Evidence for the accelerations of sea level rise on multi-decade and century timescales', *International Journal of Climatology*, vol 10, pp129–143

8
Climate of Fear: Environment, Migration and Security

Devyani Gupta

Introduction

> *Environmental degradation is already driving economic migration out of sub-Saharan Africa and onto European shores. By tackling climate change we can lessen the push factors driving immigration. If we don't tackle it, we have to brace ourselves for population shifts on a scale we have never seen before.* (former UK Foreign Secretary Margaret Beckett, 2006)

The politics of fear increasingly governs much of thought, discourse and action nowadays. Fear of the outsider has taken centre stage, whether it be visions of seemingly unending migrants threatening order in our societies, or concerns about the pressure that they will bring to our land, leading to 'environmental suicide' (McDougall, 2003). This fear is regularly recycled in visually arresting headlines and sensationalized with exaggerated if not fictitious claims. Panic-inducing statements are commonplace: 'Never have we seen immigration on this scale: we just can't cope'; 'Migrant workers importing crime, say police'; 'Mass migration is becoming a danger to our environment'; and 'Cost of water shortage: civil unrest, mass migration and economic collapse' (Rowthorn, 2007)

It is this hysteria played out in the public eye that has helped to elevate climate change on the mainstream agenda in a way that scientific evidence and academic debates on their own had so far been unable to do. After all, until fairly recently, climate change occupied the minds of many policy-makers, lobbyists, scientists and academics; but the public generally viewed it as someone else's problem. Unlike immigrants, this was not an immediate threat. It was far away both spatially and temporally and thus could be avoided.

Migration, on the other hand, has long been a clear and present danger with an arsenal of media hyperbole, government policy, legal frameworks and policing structures designed either to keep people out or make them feel unwelcome if they do manage to get in.

So what is it about climate change that has now raised our anxieties? It is my contention that along with a real concern about the potentially debilitating effects of climate change on a global scale, Western governments also have a less benevolent reason for addressing this issue – that of keeping migrants at bay. In drawing out the linkages between climate change and migration, this chapter argues for a move away from the current approach that criminalizes migrants for the sake of 'developed nations' security' to a more holistic conception that is concerned with 'human security'.

Climate change: An age of migration?

In October 2006, the Stern Review delivered a grim forecast for people around the world – climate change threatens our access to water, food production, health and use of land. And by the summer of 2007 it looked as though climate change was already hitting home: against the backdrop of raging heat in Greece and extensive flooding in the UK, newspapers declared that 'the world is warming before our eyes' (McCarthy, 2007).

The impact of climate change therefore was no longer an 'if' but a 'when and how'. The distant threat could come knocking on our doors in as little 50 years as Stern estimated that '200 million more people may become permanently displaced due to rising sea levels, heavier floods and more intense droughts' (Stern, 2006). More recently, the Intergovernmental Panel on Climate Change (IPCC) Working Group II report had a similarly bleak outlook: 'stresses such as increased drought, water shortages and riverine and coastal flooding will affect many local and regional populations. This will lead in some cases to relocation within or between countries, exacerbating conflicts and imposing migration pressures' (Schneider et al, 2007).

A number of charities and think tanks have followed suit, claiming that 1 billion people could be forced to leave their homes over the next 50 years as the effects of climate change exacerbate an already serious 'migration crisis' (Christian Aid, 2007). 'Natural disasters, together with the effects of resource stripping, have displaced millions' (Conisbee and Simms, 2003). Although there are no conclusive global figures on the number of people who are at risk of displacement, Professor Norman Myers suggests that in 1995 there were at least 25 million people displaced for environmental reasons over and above the 27 million 'traditional refugees', and 'when global warming takes hold, there could be as many as 200 million people' (Myers, 2005).

These gloomy predictions of extreme weather patterns that exert pressure on already vulnerable populations to seek refuge elsewhere have prompted alarm in the press and public alike.[1] A headline in *The Times* captured the point succinctly: 'Climate change could create 200 million refugees' (Leake, 2007). It seems that it is not fear of climate change alone – rather, the possibility that this will bring an unprecedented and overwhelming number of migrants to our gates and with them the likelihood of violence and conflict over scarce resources. Indeed, Margaret Beckett issued a stark warning to the United Nations Security Council in 2007 that 'an unstable climate risks some of the drivers of conflict – such as migratory pressures and competition for resources – getting worse' (cited in Eccleston, 2007).

Environmental exodus: Securitizing the threat of climate change

The threat of migration posed by climate change has entrenched itself within the political agenda to the extent that developed states are investing a sizeable amount of time and resources to study the risks that it poses to security. A report commissioned by the Pentagon's Office of Net Assessment (ONA) that was to prove embarrassing to the Bush administration called for climate change to be 'elevated beyond scientific debate to a US national security concern'. It also claimed that 'the impacts of climate change could potentially de-stabilize the geopolitical environment, leading to skirmishes, battles, and even war due to resource constraints' (Schwartz and Randall, 2003).

A number of events have also been dedicated to the subject,[2] such as the Royal United Services Institute for Defense and Security Studies (RUSI) conference Climate Change – the Global Security Impact in January 2007, which also explored the potential of terrorist extremists using climate change to their own advantage. Similarly, the German Federal Ministry for the Environment stated that 'environmental policy in the 21st century is also economic policy, energy policy, security policy'. Echoing these sentiments, Sir David King, the UK government's chief scientific adviser, expressed concern that 'climate change is the most severe problem we are facing today, more serious even than the threat of terrorism'.

The securitization of climate change has found expression in a number of military strategic reviews dedicated to assessing the high-risk areas around the world (Quality of Life Policy Group, 2007).

The CNA Corporation report (a panel of retired senior US military officers) points to the potential for 'racial and religious tensions' as well as 'violence between migrants and natives' as resources become scarce among overburdened communities (CNA Corporation, 2007). In a report published in 2007, the Development Concepts and Doctrine Centre (DCDC) of the UK Ministry of Defence (MoD) claimed that 'food and water insecurity will drive mass migration from some worst affected areas and the effects may be felt in more affluent regions' (DCDC, 2007). Subsequently, the MoD has announced its decision to fund a UK£12 million research contract with the Met Office Hadley Centre's climate change experts to identify regions of the world where global warming could spark conflict and security threats (Adam, 2007).

This has spawned a plethora of studies exposing the alleged linkages between climate change, migration and conflict. The German Advisory Council on Global Change (WBGU) proclaimed that climate change will trigger numerous conflicts over the management of migration and that strengthening the most vulnerable countries' adaptive capacities would 'make it easier for them to remain in their homes' (WBGU, 2007). Gledistch et al (2007) also indicate the possibility of conflict from environmentally induced migration. Meanwhile Jon Barnett is more cautious about the likelihood of such conflict, but admits that 'climate-induced conflicts are most likely as a result of migration' (Barnett, 2001).

Stemming the tide: Development instead of migration?

Conscious that climate change could have a more direct impact given its accompanying threat of an upwards trajectory in migration and the accompanying conflict, alarm bells have been ringing; consequently, politicians are paying much closer attention to 'going green'. This is partly out of genuine concern, but equally an electorally conscious practice to give the impression of reining in a situation that is perceived to be spiralling out of control.

Responding to and reinforcing these fears, therefore, UK Prime Minister Gordon Brown's first major statement to parliament in late July 2007 was to spell out a vigorous new security and border strategy. On the other hand, Douglas Alexander, the secretary of state for international development, promised that the UK would help 'to carry out research into climate change adaptation in Africa' (Alexander, 2007). As part of a UK£50 million project, the UK Department for International Development (DFID) will help '32,000 families who live on the shifting sands of the Char lands in Bangladesh ... [to] raise their homes above flood level, helping them to stay safe and build for the future'.

This proactive stance on the home front to strengthen our borders and curb immigration is positioned against a more reactive stance abroad to promote economic development and enable better adaptive capacities for environmental crises to suppress the reasons for migration in the first place (Miliband, 2007a).[3] As Sara Curran argues, the focus should be on 'the redirection of international investment in places of migrant origin to stem the flow and reverse the tide' (Curran, 2002). Much like the 'development and remittances instead of migration' policies (De Haas, 2006), the Environmental Transformation Fund – a joint venture of DFID and the UK Department for Environment, Food and Rural Affairs (Defra) – 'will support development and poverty reduction through environmental protection and will help developing countries invest in clean energy, avoid deforestation, and adapt to climate change' (De Haas, 2006). Gleditsch et al (2007) state the case more explicitly: 'because environmental disasters in developing countries may lead to greater emigration, providing generous assistance programmes will serve to limit emigration at its source'.

Activists would argue that this support is long overdue. As the biggest climate change perpetrators, it becomes our responsibility to help those most vulnerable to cope with the impacts. However, this willingness to act (at least in rhetoric) is certainly not altruistic in its sentiment – it is the careful manoeuvring that ties aid to diminish migration pressures. Giving this dubious approach credibility due to his credentials as a respected academic in this field, Professor Myers suggests that 'sustainable development represents a sound way to pre-empt the environmental refugee issue in its full scope over the long run' (Myers, 2005). Perhaps unintentionally, Myers has provided developed nations with a politically more acceptable means to prevent migrants from crossing their boundaries – a zero immigration policy with a humanitarian face.

Adaptation policies: Managing the environment for the benefit of all?

Myers (2005) argues for official recognition of what he calls 'environmental refugees' and suggests 'there would be a handsome payoff on investment to foster sustainable development in developing countries' as a means of reducing the need to migrate. This assumes that migration is a problem that requires a solution rather than understanding human movement as a basic feature of human society from prehistoric times. People always have and always will move, and projecting it as a taboo word, as increasingly many societies do, is to try to control an irrepressible and positive force.

Pointing out the futility of their cause, Saskia Sassen and others have exposed the largely failed border security policies and other so-called 'smart solutions' employed by American and West European nations in an effort to restrict immigration (Hayter, 2000; Sassen, 2006). These policies have not only been unsuccessful in their stated goals, but they have had the unintended effect of encouraging underground people trafficking, while pushing people into *permanent* settlement rather than *circular* migration patterns.

Such 'restrictivist' policies therefore reflect a misguided and incomplete understanding of migration. Distinguishing migration into neat 'economic', 'political' or 'environmental' categories is misleading and detrimental to the cause of such vulnerable people. Migrants can be all of these things and more, and the extent to which they can be pigeon holed as 'forced' or 'voluntary' is not always clear since people move for a multitude of complex and multilayered reasons. The migrant experience can be both out of compulsion (e.g. fear of political persecution as well as economic deprivation due to natural disasters and so on) and a degree of choice (although the extent to which that choice is meaningful depends upon the immediacy and scale of the threats faced by migrants).

For example, the legal understanding of the term 'refugee' as it currently stands is misguided. As Kate Romer explains, the Refugee Convention does not protect people displaced by environmental factors: 'Use of the term without any legal expansion of the definition gives governments grounds to disregard advocacy on behalf of the environmentally displaced' (Romer, 2006). Gaim Kibreab suggests that this is already happening as 'the concept of the environmental refugee is increasingly used by states to justify restrictive refugee policies' (Kibreab, 1997). This explains why Richard Black objects to the use of the term 'environmental refugee', as it not only suggests a less pressing humanitarian reason to grant asylum, but also denies individuals any legal recourse for claiming refugee status (Black, 2001).

This is not to say that people are not or could not, in future, be displaced because of climate change. There is overwhelming evidence to support the significant danger that climate change poses, especially to developing countries. For instance, scenarios where small island states risk being submerged due to sea-level rise are certainly feasible. Equally, a scenario where people escape water or food shortages because the land upon which they depend can no longer sustain them is also possible. People fleeing such peril no doubt require protection – although it should not come to a matter of life and death before they are offered the choice to migrate.

Conclusions: From 'state security' to 'human security'

> *We are frustrated, and we are angry at the same time. We are victims of something we are not responsible for... I think it's about time these industrialized countries realized that these countries in the Pacific are taking the toll. We are bearing the brunt of all the gas emissions. Millions and millions of dollars are spent on wars all over the world. Can they save people like [us]?* (Bernard Galie, resident of Piul, Carteret Islands, 2006)

The increasingly fashionable linkage of migration, climate change and security suggests that we are putting at risk something as important as the security challenges of climate change – the fact that we seem to be wholly replacing moral duty with perceived economic and strategic interest. Dubious policies that offer development assistance, with one hand, and restrict migration, with the other, should be challenged not only because they are ineffective but because this is a slippery slope that could lead to cessation of any kind of movement – even when severe and immediate dangers threaten people's lives. Faced with similar threats to their lives, would the people of developed nations expect a warm welcome from the vulnerable societies that they keep at arm's length today?

I am therefore inclined to agree with Franck Düvell who derives the following requirement for a just outcome:

> *To satisfy [moral equality and individual freedom] requires a global realization of basic rights and basic goods ... and a form of cosmopolitan membership, temporary citizenship or globally obligatory and enforceable universal rights. Both would replace the need to migrate by the free choice to migrate, and would therefore certainly also reduce migration pressure. Combined, these measures have the potential to reconcile the sedentary and the mobile and would mark the only true way to global social justice and equality.* (Düvell, 2003)

Regrettably, the practical implementation of such an outcome is likely to be beyond the political, social and economic appetite of the fortunate few in the West. However, as an initial step, the following recommendations are intended to feed debate on the issue in the hope that genuine progress can be made for those in fear of their 'human security' (Afzal, 2006):

- The understanding and application of the term 'migrant' or 'refugee' should be expanded to that which meets his/her 'human security'. This is 'an integrated, sustainable, comprehensive security from fear, conflict, ignorance, poverty, social and cultural deprivation, and hunger, resting upon positive and negative freedoms' (Van Ginkel and Newman, 2006).
- Adopting this 'human security' framework will bring fairness, coherence and efficacy to an otherwise disjointed policy set on aid, migration and climate change.
- Any development aid targeted at supporting developing countries to cope with the effects of climate change should not be tied by a reduced quota of permitted migrants or asylum seekers.
- The 'fear agenda' should be questioned and challenged so that the media and governments do not incite unhelpful and inaccurate slogans on immigrants.
- A concerted effort should be made to encourage positive and open debates on the issues without criminalizing migrants and their communities.

As David Miliband suggested 'it is neither practical nor, in the case of the developing world, morally justifiable to expect citizens to lower their aspirations and miss out on better living standards' (Miliband, 2007b). It is time we translate these words into action.

Notes

1 The most commonly cited academics linking environmental crises to mass populations movements are Homer-Dixon (1991), Myers and Kent (1995), Black (2001) and Castles (2002).

2 Other events include the joint United Nations University (UNU)–United Nations Development Programme (UNDP)–United Nations High Commissioner for Refugees (UNHCR)–International Organization for Migration (IOM) panel discussion Environmental Refugees: The Forgotten Migrants in May 2007, and the Institute for Research on Public Policy workshop on Environmental Change, Population Displacement and Acute Conflict in June 1991.

3 This is not to say that the UK government is not also motivated by humanitarian concerns. But the appeal of halting migration in its tracks appears to be as important, if not the overriding aim. As David Miliband recently explained: 'This mission is, for my government, driven by both moral purpose and national interest. We must address the insecurity and inequality that exists beyond our borders if we are to enhance security and prosperity within our borders' (Miliband, 2007a).

References

Adam, D. (2007) 'Move to identify climate change security hot spots', *The Guardian*, 11 September 2007

Afzal, M. (2006) *A Violation of His or Her Human Security – New Grounds for the Recognition of Refugee Status: A Proposal for Reform*, UNHCR Research Paper No 140, December 2006, UNHCR, Geneva

Alexander, D. (2007) 'The role of international development in a changing world: The perspective from Britain', Speech at the Council on Foreign Relations, Washington, DC, 12 July 2007

Barnett, J. (2001) *Security and Climate Change*, Tyndall Centre for Climate Change Research, Working Paper 7, UK, October 2001

Beckett, M. (2006) Speech at the British Embassy, 24 October 2006, UK Government, Berlin

Beckett, M. (2007) Speech at the UN Security Council Debate on Energy, Climate and Security, 21 April 2007, New York, NY

Black, R. (2001) *Environmental Refugees: Myth or Reality?*, UNHCR Working Paper No 34, UNHCR, Geneva, March

Castles, S. (2002) *Environmental Change and Forced Migration: Making Sense of the Debate*, UNHCR, Geneva

Christian Aid, (2007) *Human Tide: The real migration crisis*, Christian Aid, London

CNA Corporation (2007) *National Security and the threat of Climate Change*, CAN, Virginia

Conisbee, M. and A. Simms (2003) *Environmental Refugees: The Case for Recognition*, New Economics Foundation, London

Curran, S. (2002) 'Open borders and the environment: Externalities and unexpected consequences', Discussion paper for Cyber-Seminar Should Borders be Open: The Population and Environment Dimension, Population Environment Research Network, New York, NY

DCDC (Development Concepts and Doctrine Centre) (2007) *The DCDC Global Strategic Trends Programme 2007 – 2036*, Ministry of Defence, London

De Haas, H. (2006) *Turning the Tide? Why 'Development instead of Migration' Policies are Bound to Fail*, International Migration Institute Working Papers, Oxford, UK

Düvell, F. (2003) 'The globalization of migration control', www.nadir.org/nadir/initiativ/agp/campanas/migration/iomarticle.htm

Eccleston, P. (2007) 'One billion climate change refugees by 2050', *The Telegraph*, 14 May 2007

Gleditsch, N. P., R. Nordas and I. Salehyan (2007) *Climate Change and Conflict: The Migration Link*, International Peace Academy, Coping with Crisis Working Paper Series, IPA, New York, NY

Hayter, T. (2000) *Open Borders – The Case Against Immigration Control*, Pluto Press, London

Homer-Dixon, T. (1991) 'On the threshold: Environmental changes as causes of acute conflicts', *International Security*, vol 16, pp76–116

Kibreab, G. (1997) 'Environmental causes and impact of refugee movements: A critique of the current debate', *Disasters*, vol 21, no 1, pp20–38

King, D. (2004) http://news.bbc.co.uk/1/hi/uk_politics/3584679.stm

Leake, J. (2007) 'Climate change "could create 200m refugees"', *The Times*, 1 April 2007

McCarthy, M. (2007) 'The world is warming before our eyes', *The Independent*, 28 September 2007

McDougall, R. (2003) 'Mass immigration: A route to environmental collapse', Submitted to Open Democracy debate on the Demos report *People Flow*, www.opendemocracy.net/people-migrationeurope/article_1233.jsp

Miliband, D. (2007a) Speech at the UN General Assembly, New York, 27 September 2007

Miliband, D. (2007b) Speech at the WWF One World Living Summit, London, 15 March 2007

Myers, N. (2005) 'Environmental refugees: An emergent security issue', Speech at the 13th Economic Forum, Session III Environment and Migration, Prague, 22 May 2005

Myers, N. and J. Kent (1995) *Environment Exodus: An Emergent Crisis in the Global Arena*, The Climate Institute, Washington, DC

Quality of Life Policy Group (2007) *Blueprint for a Green Economy*, www.qualityoflifechallenge.com/

Romer, K. (2006) 'Environmental refugees?', *Forced Migration Review*, no 25, May 2006

Rowthorn, R. (2007) 'Never have we seen immigration on this scale: We just can't cope', *The Daily Telegraph*, 2 July 2007

Sassen, S. (2006) 'Migration policy: From control to governance', Submitted to Open Democracy debate on the Demos report *People Flow*, London

Schneider, S. H., S. Semenov, A. Patwardhan, I. Burton, C. H. D. Magadza, M. Oppenheimer, A. B. Pittock, A. Rahman, J. B. Smith, A. Suarez and F. Yamin (2007) 'Assessing key vulnerabilities and the risk from climate change', in *Climate Change 2007: Impacts, Adaptation and Vulnerability. Contribution of Working Group II to the Fourth Assessment Report of the Intergovernmental Panel on Climate Change*, IPCC, Cambridge University Press, Cambridge, UK, pp779–810

Schwartz, P. and D. Randall (2003) *An Abrupt Climate Change Scenario and Its Implications for US National Security*, Global Business Network, San Francisco, CA

Stern, N. (2006) *The Economics of Climate Change*, HM Treasury and Cabinet Office, Cambridge University Press, Cambridge, UK

Van Ginkel, H. and E. Newman (2006) cited in M. Afzal (2006) *A Violation of His or Her Human Security – New Grounds for the Recognition of Refugee Status: A Proposal for Reform*, UNHCR Research Paper No 140, December 2006, UNHCR, Geneva

WBGU (German Advisory Council on Global Change) (2007) *World in Transition: Climate Change as a Security Risk*, WBGU, Berlin

9
Changing Economics

Nicholas Stern

Introduction

Managing climate change and overcoming poverty are defining challenges of the 21st century. We will succeed or fail on the two together. If we fail to manage climate change, we will create a physical environment so hostile it will destroy the prospects of development for billions of people. On the other hand, if we try to manage climate change in a way that obstructs, or appears to obstruct, prospects for development for poor countries over the next few decades, we will not create the coalition necessary to combat climate change. We can and we must tackle these two challenges together.

But there is a third issue. The world is currently emerging from a deep global economic and financial crisis, the result of inadequate risk management within the financial sector, regulatory shortcomings and global macro-economic mismanagement. The scale of risk from climate change is altogether of a different and greater magnitude than the financial crisis, as are the consequences of mismanaging or ignoring it. Some may argue that the current economic and financial circumstances should lead to the postponement of action on climate change. That would be a serious mistake. A key lesson of the present financial crisis should surely be that it is dangerous to ignore, or fail to recognize, the build-up of risk: this current economic crisis was 15 or 20 years in the making. If the world postpones action on climate change for a similar period, our starting point will be much more difficult and the risks faced will be far greater.

While this third crisis is of a smaller magnitude than the other two, it is major and urgent and must be tackled together with them. With good policy we can, indeed, tackle all three together. Although this chapter is focused on the economic analysis of policy for climate change, the measures it proposes combine climate change and fighting poverty, and should be at the heart of green fiscal measures to combat the economic recession or slow down.

Strong leadership is required to drive the necessary action. In many countries, evidence of this leadership is now emerging through a renewed focus on climate change and recognition of the potential for sustainable low-carbon growth. Continued public discussion and pressure for action will be critical.

This chapter is based on the findings of the Stern Review (Stern, 2007), subsequent publications, and interactions with academics and policy-makers on the issue of climate change. The chapter will discuss the problems and related impacts of climate change. It examines the economics in terms of the risks, the costs of action, appropriate targets, the relevant policy instruments and the challenge of adaptation. The chapter concludes with a description and analysis of the key elements of a global deal that could be negotiated at the United Nations Conference in Copenhagen in December 2009, with respect to the criterion of effectiveness, efficiency and equity.

The problem of climate change

The flows of greenhouse gases (GHGs) that human activity causes are larger than the planet can absorb. Thus, concentrations of GHGs in the atmosphere rise. These extra concentrations trap more heat, causing global warming. The global warming causes climate change. And climate change disrupts human activity. In the language of economics, this damage to others from emissions is an externality.

Climate change is the greatest market failure the world has ever seen. GHGs represent an externality that is different from most examples in economics in four fundamental ways. First, this externality is global; 1 tonne of CO_2 emitted in the US has a similar social impact as 1 tonne of CO_2 emitted in the UK. Second, emissions and the accumulation of stocks in the atmosphere have long-lasting impacts of up to a century or more. Third, the science points to inherent risks and uncertainties regarding the extent and impacts of climate change. Last, the effects of this externality are potentially very large and may not be reversible. Therefore, these features profoundly influence the relevant economic policy.

Economic policy analysis must be based on risk and uncertainty, it must consider ethics, including intergenerational concerns, and it must be global in its policy prescriptions. Failure to analyse the problem in terms of the scale of the global risks, the long term and through an international policy perspective will, and has, led to economic analysis and policy that is misleading and dangerous.

While some of the much loved traditional economic tools of marginal analysis remain useful in is context, it is clear that this approach is inadequate to guide an effective policy response on its own. The economic analysis must consider non-marginal changes in economic structures and examine radically different growth paths in the presence of significant uncertainty and risk.

People around the world are suffering from past emissions. The affects of an approximate 0.7°C increase in global average temperatures relative to pre-industrial times (mid 19th century) are already evident: floods in Mozambique in 2000 that decreased gross domestic product (GDP) by at least 10 per cent; movement of pastoral populations in Darfur as a result of extended drought, causing conflict with those in settled agriculture; retreat of glaciers in the Himalayas affecting water supplies throughout South and East Asia; heat waves of 2003 in Europe; hurricanes in the US and the Caribbean of increased frequency and intensity. Looking to the future, even if the world acts responsibly, the planet is likely to see a further 1°C to 2°C increase in average global temperatures. If, however, the world continues with business as usual (BAU), the world has around a 50–50 probability of a temperature increase exceeding 5°C above the mid 19th century (Table 9.1).

A temperature change of this magnitude would transform the planet. Only 10,000 to 12,000 years ago, when temperatures were 5°C lower than today, most of Northern Europe and North America were under hundreds of metres of ice and the distribution of human populations was restricted to areas closer to the equator. A continued failure to take responsibility for our future will result in tremendous movements of population and hundreds of millions, possibly billions, will have to move, with the likelihood of severe and protracted conflicts. While the costs of disruption on this scale are very hard to quantify, a rough calculation, presented in the Stern Review, suggests that the world may experience a cost, averaged over space, time and possible outcomes, equivalent to 20 per cent of annual consumption every year. If, however, the world takes responsibility now, the risk from climate change can probably be significantly reduced at a cost of roughly 1 to 2 per cent of GDP (between –1 and 3 per cent). Thus, the cost of strong and timely action is much less than inaction or delayed action. A decisive and committed response makes economic sense. Furthermore, as we shall argue, climate responsibility can be combined with economic growth and development. This is not a horse race between economic growth and development, on the one hand, and climate responsibility, on the other. Climate responsibility and low-carbon growth together constitute the only development and growth story of the 21st century. High-carbon growth will kill itself: first from high hydrocarbon prices and

Table 9.1 *Probabilities of exceeding a temperature increase at equilibrium (percentage)*

Stabilization level (in ppm CO_2e)	*2°C*	*3°C*	*4°C*	*5°C*	*6°C*	*7°C*
450	78	18	3	1	0	0
500	96	44	11	3	1	0
550	99	69	24	7	2	1
650	100	94	58	24	9	4
750	100	99	82	47	22	9

Source: adapted from Stern (2007, Box 8.1, p220)

second, and more fundamentally, from the hostile physical environment that it would generate.

Risks

The relation between the stock of GHGs in the atmosphere and the resulting temperature increase is at the heart of any risk analysis and discussion of the economics. It is the clearest way to begin and anchors most of the discussion. While the greenhouse effect was discovered and its causes identified during the 19th century, science has in recent years begun to provide us with probability distributions of temperature outcomes for different stocks of GHGs.

Current concentrations of GHGs are around 430 parts per million (ppm) CO_2 equivalent (CO_2e) – which aggregates carbon dioxide with other GHGs – and are increasing by around 2.5ppm CO_2e per year. This rate appears to be accelerating, particularly as a result of rapid growth of emissions in the developing world. There seems little doubt that, under BAU (i.e. in the absence of any restraining policy), the annual increase in the overall quantity of GHGs would average somewhere above 3ppm CO_2e, potentially 4ppm CO_2e or more, over the next 100 years. This is likely to result in concentrations of around 750ppm CO_2e by the end of this century.

This level of concentration, if there were no further growth, would result in a 50–50 chance of a temperature increase over 5°C. In contrast to when temperatures were 5°C below pre-industrial levels, the planet last experienced temperatures 5°C above pre-industrial levels around 30 million to 50 million years ago, long before the arrival of humans. The most recent warm period was around 3 million years ago when the world experienced temperatures 2°C or 3°C higher than today (Jansen et al, 2007). Humans have never experienced such high temperatures. Many areas, such as much of Bangladesh, would be inundated; others such as Southern Europe would become deserts; rivers would change course; hurricanes would batter the southern US. The physical and, thus, human geography (where we can live and how we live our lives) would be transformed.

Targets

In order to reduce the risk of climate change, the world must act together and commit to targets for emissions reductions. Current evidence suggests the world should try to hold concentrations below 500ppm CO_2e and then try to reduce from there. This would involve emissions peaking in the next ten years and reducing to half 1990 levels, or 20 gigatonnes (Gt) CO_2e at most by 2050. This would be equivalent to around 2 tonnes per capita in 2050. Thus, the global targets discussed at the Heiligendamm G8 summit in 2007 and confirmed at the 2008 Hokkaido Summit are encouraging and illustrate that the world is beginning to talk about the necessary levels of commitments required.

Rich countries, with their past responsibility for emissions, better access to technology and better access to resources should commit to at least an 80 per cent reduction in emissions by 2050. This would bring European emissions down to around 2 tonnes per capita and corresponds to the commitment made by the UK government to reduce its emissions by 80 per cent on 1990 levels by 2050. President Obama has indicated an 80 per cent target for reductions for the US (1990 to 2050), although it should be recognized that since US emissions per capita are currently more than 20 tonnes CO_2e, or around twice those of Europe, this would leave the US around 4 tonnes per capita in 2050.

In addition to these distant objectives, intermediate targets for 2020 are necessary. Intermediate targets stress the immediacy of the problem, increase the incentive to take early and strong action and show credibility for other countries, businesses and markets. The 20 to 30 per cent reductions for 1990 to 2020, as set by the European Union (EU) Council in March 2006, are a good guide for the rest of the world. If 80 per cent reductions are to be achieved in the 60 years of 1990 to 2050, then more than 20 per cent at 'halfway' are surely necessary.

Cost of abatement

Different emissions paths involve different risks and costs.

Figure 9.1 illustrates possible paths that could stabilize at 550ppm CO_2e (second from top), 500ppm CO_2e and 450ppm CO_2e (bottom line); the top black line is BAU. The 500ppm path could, if emissions kept on falling, eventually result in stabilization around 450ppm CO_2e. Possible paths for stabilization at a given level are presented in the Stern Review (Stern, 2007, Figure 8.2, p226) – but all of them are a similar shape to those shown (if a path peaks later it has to fall faster). And if the carbon cycle weakens, the cuts would have to be larger to achieve stabilization at a given level (see Stern, 2007,

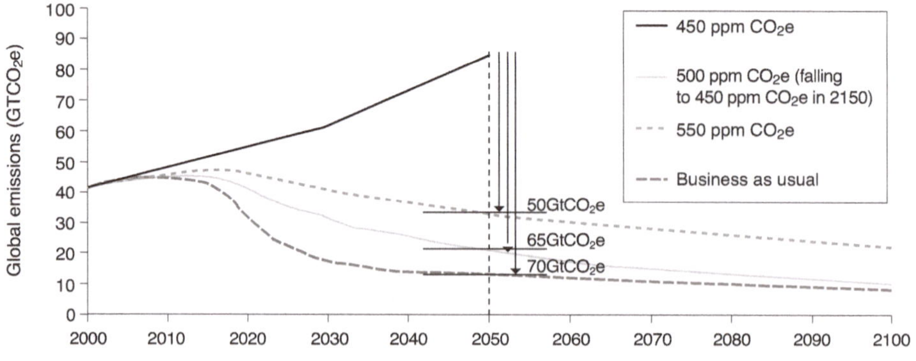

Source: adapted from Stern (2007, Figure 8.4, p233)

Figure 9.1 *Possible paths for the stabilization of greenhouse gases (GHGs)*

Figure 8.1, p222). Broadly speaking, however, a path stabilizing at 550ppm CO_2e or below will have to show emissions peaking in the next 20 years. For lower stabilization levels, the peak will have to be sooner. The magnitudes of the implied reductions between 2000 and 2050 are around 30 per cent for 550ppm CO_2e, 50 per cent for 500ppm CO_2e, and 70 per cent for 450ppm CO_2e. Cuts relative to BAU are indicated in the figure.

On any emissions reduction path there are different mitigation options. At the start of the stabilization path there is great scope for energy efficiency and measures to slow/halt deforestation; with technical progress there will be (and some are there already) strong roles for different technologies in power and transport.

The cost of achieving these stabilization levels, if the world acts now, is relatively small, compared to the damages avoided. Both the bottom-up and the top-down studies in the Stern Review (Stern, 2007, Chapters 9 and 10, respectively) produced costs in similar ranges – around 1 per cent (between –1 and 3 per cent) of world GDP for stabilization below 550ppm CO_2e. With hindsight and ever clearer evidence, 550ppm CO_2e would be too high as a target: the risks look bigger than we estimated for the Stern Review. Concentrations should be held below 500ppm CO_2e and, if possible, eventually brought down to no higher than 450ppm CO_2e. The annual costs over the next few decades might be around 2 per cent of world GDP.

Bad policy, delayed decisions or incomplete participation could result in significantly higher costs. Stronger technical progress than anticipated could lead to lower costs. Assumptions about substitutability between different goods and options matter too. Since the Stern Review was published there have been a number of new studies, both bottom up and top down. Significant examples of the former are those from McKinsey (Enkvist et al, 2007) and the International Energy Agency (IEA, 2007), both of which indicated costs consistent or lower than the Stern Review. Similar conclusions on costs of action are drawn in the *Fourth Assessment Report* of the Intergovernmental Panel on Climate Change (IPCC, 2007).

Adaptation

Climate change affects us all. For example, in the next few decades London will have to make considerable investments in its future, ranging from the building of stronger flood defences, larger sewer systems and the provision of air conditioning for the London Underground. This is as a result of past emissions and those which will occur in the coming few years. However, the challenges in poor countries will be much more severe than those faced by rich countries. The great inequity of climate change is that those least responsible for the present stocks of carbon in the atmosphere are those hit the hardest. Climate change has the potential to threaten all aspects of the development agenda, potentially increasing poverty and hunger levels, creating more health threats, and forcing populations to migrate and relocate. In some countries, as we have described above, some of the effects are already visible.

As global temperatures increase, all countries will need to adapt to limit the human, economic and social impacts of climate change. Even the most aggressive stabilization targets being discussed imply acceptance of a global average temperature rise of 1°C to 3°C. This will lead both to more frequent and severe climate-related disasters (including droughts, flooding and storms) and to longer-term stresses (including changing rainfall patterns, ecosystem degradation, reduced biodiversity and higher sea levels).

These changes will affect poorer countries disproportionately: not only are they typically reliant on climate-sensitive industries such as agriculture and forestry, but poverty, poor health and limited capacity and resources to adapt increase their vulnerability.

The best form of adaptation is development itself: it leads to more diversified economies, stronger and more resilient human capital, and greater flexibility in the population. And adaptation should be seen as development in a more hostile climate. Thus, it is a profound mistake to see adaptation as somehow a different activity from development.

Just as adaptation planning needs to be integrated within development plans and strategies, so adaptation funding should be integrated within development spending at regional, national and local levels. This applies to overseas development assistance (ODA) as well. ODA should be enhanced for adaptation and new parallel processes should be avoided. Support should be for national development plans, reflecting overall national priorities. This would keep down both transaction costs imposed on developing countries (by reducing the range of different institutions, people and systems governments need to deal with) and the financial costs of creating and managing a new international mechanism.

The extra costs that climate change will imply for reaching development goals are not small. The United Nations Development Programme (UNDP) in its *Human Development Report* published in late 2007 estimated an extra US$86 billion per annum (by 2015) to reach the Millennium Development Goals (UNDP, 2007). By 2020, this would be likely to be of the order of an extra US$100 billion (and that is roughly the level of ODA today). This increases the importance of delivery on the promises of aid from rich countries at Monterrey in 2002 and Gleneagles in 2005. The arguments for 0.7 per cent of GNP as development aid from rich countries by 2015 were very powerful when they were made in 2002 and 2005; in the context of climate change, they are overwhelming. The developed world must recognize that the development agenda is now much more costly than anticipated 10 to 15 years ago and raise the target to around 1 per cent of GNP per annum.

A dedicated source of additional finance would provide confidence for poorer countries that resources for the extra costs adaptation requires would be both predictable and additional to other aid commitments. As nations and regions move towards cap-and-trade systems, auctioning of emissions allowances stands to generate substantial new revenue streams, as would revenues from carbon taxes on, for example, aviation. Diverting some of this revenue to finance adaptation in developing countries would provide some reliability for developing countries and enhance the acceptability in developed countries of these measures.

Adaptation, like development itself, requires economy-wide planning and regional cooperation and therefore requires the involvement of all parts of government. While environment ministries will play a crucial role, the challenge of adaptation must be taken on by heads of state and their finance and economic ministries. There is also a role for international financial institutions, including the World Bank and the International Monetary Fund (IMF), to monitor, report on and, where necessary, facilitate non-financial aid such as access to insurance, technology, information and other market-based facilities. Disaster relief and new more climate-resistant crops and technologies will require action at international level.

A global deal

The United Nations Conference of the Parties in Copenhagen in late 2009 will be decisive in determining the post-2012 polices. It is the most important international gathering since World War II. The risks that it must grapple with and the politics it adopts must be truly global. The devastation that climate change can cause are still higher than those two world wars and the Great Depression, which prompted the establishment of the Bretton Woods institutions. Delay in coming to an agreement would be dangerous. First, the relentless flow-stock process would take us into ever more difficult territory and, second, market confidence, crucial for the necessary investment, would be undermined.

It is important that the agreement be guided by clear principles based on rigorous analytic foundations and a common understanding of the key challenges. The following describes an outline of a possible global deal based on the preceding analysis and on intensive experience and public discussion.

Any global deal must contain three basic principles:

1. *Effectiveness* – it must lead to cuts in GHGs on the scale required to keep the risks from climate change at acceptable levels.
2. *Efficiency* – it must be implemented in the most cost-effective way, with mitigation being undertaken where it is cheapest.
3. *Equity* – it must take account of the fact that it is poor countries that are often hit earliest and hardest, while rich countries have a particular responsibility for past emissions.

Effective action requires:

- Global emissions to fall by at least 50 per cent relative to 1990 levels by 2050. Emissions will need to decline to less than 20Gt per annum in 2050 and to less than 10Gt by the end of the century to hold concentrations below 500ppm CO_2e and allow eventual stabilization around 450ppm CO_2e.

- Global average per capita emissions that will, as a matter of basic arithmetic, need to be around 2 tonnes by 2050 (20Gt divided by 9 billion people). This figure is so low that there is little scope for any large group to have emissions above it (since it would be difficult to be below).[1]
- Agreement by developed countries to take on immediate and binding national targets of 20 to 40 per cent by 2020, and to commit to reductions of at least 80 per cent by 2050. Where past policy neglect has made the 2020 target very different, there could be strong contribution to reductions elsewhere and strong targets for 2025 and 2030.
- By 2020, demonstration by developed countries that they can deliver credible reductions, without threatening growth, and that they can design mechanisms and institutions to transfer funds and technologies to developing countries.
- Subject to this, a formal expectation that developing countries would also be expected to take on binding national targets of their own by 2020, but benefit from one-sided selling of emissions credits in the interim. This means rewards for reductions but not necessarily penalties for rises.
- Fast-growing middle-income developing countries with higher incomes will need to take immediate action in order to stabilize and reverse emissions growth, including sectoral targets and, possibly, earlier national targets.
- A commitment by all countries, regardless of targets, to develop the institutions, data and monitoring capabilities, and policies to avoid high GHG infrastructural lock-in.

These are essentially the requirements for 'effectiveness'.

Let us turn to efficiency and equity. Only sound, measured and coordinated policy and timely international collaboration can deliver strong and clean growth for all at reasonable cost. No strategy that requires a slowdown in the fight against poverty would be either equitable or politically viable. But it must also be clearly recognized that weak or delayed action will eventually choke off growth and be a far more costly option.

Backed by strong developed country targets for reductions, carbon prices can be maintained at levels that will provide incentives both for reductions at home and purchases from abroad. The cheapest mitigation options often reside in developing countries, which should take advantage of carbon markets from the outset. The current structure of the Clean Development Mechanism (CDM) makes it difficult to create market flows to developing countries on the scale required. Moving from a project-based to a wholesale mechanism, perhaps based on sector-specific efficiency targets and credible sector decarbonization plans, would permit scaling-up in a number of emissions and energy-intensive industries. Standardized international benchmarks on emissions-output relationships or technologies would help to reduce the risk of emissions diversion and relocation, and to alleviate competitiveness concerns in internationally traded sectors.

By putting an appropriate price on carbon, policy-makers will oblige consumers and producers to face up to the full social cost of their emissions.

Economic efficiency points to the advantages of a broadly comparable global price and coordinated policy based on carbon trading, with openness to international trade so that emissions reductions take place wherever they are cheapest. It is possible to put a price on carbon, explicitly through tax or trading, or implicitly through regulation.

It is important to weigh up the competitiveness risks and opportunities for firms, countries and sectors, especially where some countries or sectors apply GHG policies earlier and more ambitiously than others. There will be losers, and the impacts of transition will need to be managed. However, transition to a GHG-constrained world will create opportunities for companies and sectors that anticipate new markets. Moreover, the evidence to date suggests that few firms are likely to relocate activities to less restrictive jurisdictions. Overstating the problems relative to the opportunities risks prompting parties to wait for others to move before taking action. By contrast, the expectation of a credible global agreement would sharpen the incentives for companies and governments to move quickly and efficiently.

The key areas for action are deforestation, energy efficiency and low-carbon technologies. Reducing emissions from deforestation and degradation is a potentially cost-effective method of limiting emissions. Indeed, rapid results here are absolutely crucial to achieving reductions on the scale required. Furthermore, they can yield significant benefits in terms of biodiversity, watershed management and local livelihoods. Policies must be framed by the countries where the trees stand and set in the context of, and consistent with, their own circumstances and plans for development. But strong international support is essential: tropical deforestation is an international problem needing urgent international action. Addressing deforestation requires large-scale public resources and must form part of a development strategy that raises agricultural productivity, creates alternative opportunities and improves governance. There should be a long-term aim of integration within carbon markets so that private flows can take an increasingly large share.

Options for energy efficiency are everywhere and many firms and sectors have found that strong savings are immediately available if there is a clear focus on the issue. Others depend upon investments, such as insulation, many of which show high returns. Technological progress in energy efficiency (e.g. in light bulbs) is moving quickly.

For emissions to have been reduced to around 2 tonnes per capita in 2050, most of the world's electricity production will need to have been de-carbonized, while emissions from transport, land use, buildings and industry will need to have been cut sharply. The importance of technological innovation in delivering this transformation can hardly be overstated. Some cost-effective emissions reductions can be undertaken immediately using known technologies (e.g. in energy generation and transmission). In the medium to longer term, however, the task is to deliver next-generation low-carbon technologies, especially for the power, transport, industry and building sectors. Different policy frameworks will be required for different technologies at different stages of development. This will require a major scale-up in public research and development (R&D) on a global basis, support for demonstration projects, global efficiency or emissions standards, and new public–private partnerships to share risk efficiently. It is particularly important that a systematic plan for the development of carbon capture and storage technologies is developed quickly. Efficient technology policy requires globally coordinated action to pool risks and rewards, exploit economies of scale and avoid duplication. Early action to develop and deploy technologies enhances the gains from learning and experience, and promotes cost reductions through induced innovation. In addition to progressively tougher targets and a global cap-and-trade regime, any global policy framework should also aim to expand the market for low-carbon technology. Above all, it is vital that as technologies are developed they are very quickly shared across countries.

Implementation must be structured, phased and managed. A credible global institutional structure is essential in order to manage the proposals and the various different steps outlined in this chapter. Institutions need to be able to match the scale of the challenge and build trust between members, while being flexible enough to adjust to changing circumstances. Any new structure should build on the expertise of existing institutions and more detailed policy and micro-economic analysis will be required to underpin an effective, efficient and equitable global deal in Copenhagen in 2009. But the first task is to create an agreement. The institutional structures should be created around the functions necessary to implement the agreement.

An equitable agreement requires strong support for the development of poor countries. They have been least responsible for the past accumulation of GHGs and are hit earliest. This means finance for their shift to low-carbon growth. Much of this can come from carbon markets. But of great importance, too, is funding for adaptation. It is simply more costly to pursue development goals in a more hostile climate. This finance should be centred on national development plans embodying low-carbon growth and adaptation. These can only be constructed by the developing countries: they set their own goals and understand their own circumstances. But there is much that others can do beyond the vital finance. In particular, the sharing of technologies and know-how should be at centre stage. This means first delivering on existing commitments on aid and then going beyond to take account of the extra costs of a more hostile climate and the medium-term investments in low-carbon technologies.

Conclusions

If the world allows climate responsibility and growth and development to become set against each other, the argument is lost from the start. The world has both the technology and the economic understanding to move forward strongly on both simultaneously. Indeed, if it fails on one, it fails on the other.

Leadership must come from the top and from heads of state. Actions on both mitigation and adaptation must look across the entire economy from cities to tax systems. Therefore, the problem of climate change is too wide to be relegated only to one or two individual government departments or ministries. Indeed, when it comes to making the global deal it must be the heads of governments who commit to action. Therefore, it is vital that united and decisive leadership is displayed by world leaders. The statements by presidents and prime ministers alike that the investment in this financial crisis should be around green technology are to be welcomed. Encouragingly, some stimulus investment packages announced to date, with notable examples of Korea and China, contain substantial sustainable investment measures. This suggests that greener, more sustainable growth is being recognized as the only type of growth for the 21st century.

Action is not led only from the top: it will be the individual understanding of citizens, of non-governmental organizations (NGOs) and of communities that will drive forward this debate. It is striking, notwithstanding the economic crisis, that in national political debates around the world climate change is high on the agenda. This is now an electoral issue; people are pressing their leaders to act. The understanding and demands of members of the public are the most fundamental drivers of political change. It will be this voice that will carry us through to a more responsible future.

Note

1 We should note that this applies to actual emissions and there are strong arguments for rich countries to not only limit their actual emissions to these levels but also to fund emissions reductions elsewhere.

References

Enkvist, P.-A., T. Nauclér and J. Rosander (2007) 'A cost curve for greenhouse gas reduction', *The McKinsey Quarterly*, vol 1, pp35–45

IEA (International Energy Agency) (2007) *World Energy Outlook 2007*, IEA, Paris, France

IPCC (Intergovernmental Panel on Climate Change) (2007) *Climate Change 2007: The Physical Science Basis*, Cambridge University Press, Cambridge, UK

Jansen, E., J. Overpeck, K. R. Briffa, J .C. Duplessy, F. Joos, V. Masson-Delmotte, D. Olago et al (2007) 'Palaeoclimate', in S. Solomon, D. Qin, M. Manning, Z. Chen, M. Marquis, K. B. Avery, M. Tignor and H. L. Miller (eds) *Climate Change 2007: The Physical Science Basis*, Cambridge University Press, Cambridge, UK, pp433–498

Stern, N. (2007) *The Economics of Climate Change: The Stern Review*, Cambridge University Press, Cambridge, UK

UNDP (United Nations Development Programme) (2007) *Human Development Report 2007/2008: Fighting Climate Change – Human Solidarity in a Divided World*, Palgrave Macmillan, New York, NY

10

Water Scenarios: 'The Unpredictable Flow'

Johan Kuylenstierna, Michael Moore and Dave Trouba

> *The world is full of uncertainty and it is difficult to predict the future. By working together, however, we can shape the world and make it a better place – for people, for life.* (UN, 2003)

Water: The abundant resource or a planner's headache?

Water, so critical to life on our planet, is everywhere, but unevenly so. Its fickle nature – water falls and flows when and where it chooses – hinders long-term planning for its availability and use. Globally, the water cycle provides about 40,000 cubic kilometres of renewable fresh water annually (Shiklomanov, 2000); but such a number is of little value to the water resources planner in a country or at the local level since the differences in rainfall, evaporation (from land and water) and transpiration (from plants)[1] vary greatly by region and over time. Some regions experience heavy precipitation for parts of the year followed by long stretches of near total dryness (monsoonal and some temperate climates), while others such as Africa's Sahel region have large multi-annual or multi-decadal variability. Only some parts of the temperate zone enjoy fairly stable and reliable precipitation and benefit from low evapotranspiration.

The basin, national or local scales are more relevant for water resources planning. The water available in a basin (whether rain or water in a river, the soil or underground) is what humans and nature can access and use. Understanding the seasonal, annual and long-term variability at the basin scale is therefore essential for good management and planning.

In the basin, water's mobility challenges us. Water, both surface and groundwater, moves continuously through the basin landscape, although for planners, reality is different. Water boundaries do not define political borders; yet water availability in a country or a part of a country is reported according to the political boundaries. Thus, for most planners it is the city, district, county, provincial or national political divisions that serve as the starting point for most water management policies and strategies.

Complicating this is that water is often shared by countries. Rain is the dominant water source in some countries (in the temperate zone, or the most upstream country in a river basin); but for others it is the inflow of water from upstream countries that dominates (Egypt receives 97 per cent of its available water resources from upstream countries in the Nile Basin). This transboundary dimension greatly affects the degree of control and freedom of use in planning.

Global water use trends: Ever increasing?

Water is fundamental for all kinds of development. Human development and well-being, industrial and economic development, energy production and agriculture all rely on the availability of (often abundant) water resources. However, there are distinct differences between sectors of society, their relationship to water and the future implications that changes in availability may have.

The 20th century witnessed a sixfold increase in human water use, while the population increased 'only' threefold (Cosgrove and Rijsberman, 2000; Gleick, 2003). It is not only direct human water use that influences the quantity of water; availability could also be indirectly affected through water quality deterioration due to increased pollution and deteriorated aquatic ecosystems (Koplin et al, 2002; Baresel and Destouni, 2005; Lindgren et al, 2007). Water use statistics, should be treated carefully as they may include large uncertainties. As an example, we currently do not know if groundwater provides only 25 per cent or as much as 40 per cent of the world's drinking water (UN, 2006).

So, what is water used for?

In our houses

Domestic water use (drinking, sanitation and societal services) accounts for approximately 8 per cent of global water use, though at the household level it varies considerably by one's living standard (UN, 2003). In 2000 nearly 1 billion urban dwellers lived in slums and had as little as 5 litres per day at their disposal, while nearby middle- or high-income households had between 50 to 150 litres per day (UN, 2006). In 2006, according to the latest Joint Monitoring Programme report, 884 million people lack access to safe drinking water (WHO and UNICEF, 2008). This is nothing less than a scandal.

To produce our goods and services

Industry is a thirsty water consumer and accounts for some 22 per cent of total use (UN, 2006), but with great regional variation. In rich countries, industry can command as much as 59 per cent of the water while developing countries dedicate roughly 8 per cent of their available water. Global economic growth has and will continue to fuel an expansion of industrial water use from about

750 cubic kilometres per year in 1995 to an estimated 1170 cubic kilometres per year in 2025. In Asia, for example, water withdrawal is five times higher today than in 1950 (Shiklomanov, 2007).

Industry is also a major water polluter. Some 300 million to 500 million tonnes of heavy metals, solvents, toxic sludge and other wastes accumulate each year from industry, 80 per cent of which is produced in industrial countries, but as much as 70 per cent of which are dumped untreated into waters in developing countries (UN, 2006). Therefore, industrial water use should also include the amount of water degenerated due to pollution; statistics, however, are lacking.

Predicting industry's future water use is difficult. Technologies can cut water use dramatically (in some cases to zero); but clear economic and legal incentives need to promote such technologies. Which industries grow is also a critical factor due to differences in water use and pollution generation by sectors.

To keep society running

Energy production is also a major water issue. However, the linkages are not always obvious. Reservoirs would seem to be built for energy generation, but merely 25 per cent of the world's dams produce hydropower. Europe uses 75 per cent of its hydropower potential, while Africa has developed only 7 per cent, a figure so low that raising it is viewed by some to be a cornerstone of Africa's future development (UN, 2006). Still, only about 2.2 per cent of the total primary energy supply by source (in 2002) comes from hydropower, according to the International Energy Agency (IEA), and even if there is a dramatic increase in hydropower production, the overall share will remain marginal.

Water is also used as a coolant in energy production with little effect on quantity; but the quality and temperature of the water may be affected in the process. Increased water use, changes in runoff due to climate change and higher water temperatures may similarly decrease the cooling capacity of a specific water source. Such factors will have to be considered where massive energy development (including coal-powered plants) will need to use huge amounts of water in fast growing economies such as China and India.

A debate emerging in recent years has centred on the water needs for bioenergy production. A dramatic production increase could drastically alter future water use – and food production – scenarios. The energy section of the 2006 *World Water Development Report* (UN, 2006) barely addressed bioenergy and water linkages. With some estimating that as much additional water is needed to meet our bioenergy needs as to meet our food needs in the future, this issue will only grow in importance in the global water debate.

To feed us

At 70 per cent of global water use, the thirstiest sector clearly is, and will remain, agriculture, which has tripled its irrigated area between 1950 and 2003 from 94 million to 277 million hectares (producing 60 per cent of the world's grain). This irrigation growth has tapered as water has become scarcer. Forty years ago, irrigated areas were expanding at an annual rate of 2.1 per cent; but the last five years show growth of only 0.4 per cent (FAO, 2006).

Agricultural water management has also seen dramatic changes since the 1950s. Advances in seed, fertilizer and pesticide technologies, and the ability to store, divert and pump surface and groundwater have spawned 'green revolutions' nationally. Dams, diversions and other infrastructure harnessed blue water (lake, river and groundwater) resources for farming, hydropower and flood protection. As a result, it is estimated that 60 per cent of the large river systems in the world have been moderately or strongly affected (MEA, 2005a). Since 1950, water withdrawals have tripled, although the resource has not been used or managed efficiently. Globally, food production has kept pace with a doubling of the world population, a yield increase due more to increased cropping intensities than to expansion into previously uncultivated areas (Molden, 2007).

Water use in agriculture, whether for livestock husbandry, aquaculture or crop production, has also affected water quality. Soil erosion, salinization, sedimentation, and nutrient and pesticide pollution have contributed to a decline in water quality in many of the world's river systems and coastal zones.

The net result of population growth, economic development and improved diets is a drastic increase in water use, thus increasing risks of competition. Agriculture, accustomed to receiving the lion's share of water, but often showing the lowest economic return per drop, is being pressured by industries, cities and ecosystems, which need water to maintain their productive yet often undervalued services (UN, 2006). However, the true costs of these advances are only starting to be known. ´The lack of appropriate and efficient water resources management in the agricultural, industrial and domestic sectors have all contributed to degradation of ecosystems (Molden, 2007). The Millennium Ecosystem Assessment (MEA, 2005a) stated that humans have changed ecosystems more rapidly and extensively than ever before in the last 50 years in order to meet our growing demands for food, freshwater, timber, fibre and fuel.

The main drivers for the degradation of freshwater ecosystems have been habitat alteration through wetland drainage and land-use change, alteration of river flows and pollution. Climate change is also expected to increase the pressure on freshwater ecosystems (UN, 2006). The withdrawal of water for human use, and particularly for irrigation, is contributing to a shift to environmentally stressed conditions in many river basins around the world (Smakhtin et al, 2004). There are an increasing number of cases around the world where rivers are no longer reaching the sea and groundwater tables are dropping as we are overusing our global freshwater by 5 to 25 per cent more than the long-term accessible supply allows. Much of this water is used for irrigation, with irretrievable losses in water-scarce regions, although overuse is everywhere (MEA, 2005b). Huge demands are also being made on coastal and freshwater ecosystems, where freshwater species are threatened by human activities. Species populations fell 50 per cent between 1970 and 2000, representing a sharper decline than measured in either terrestrial or marine biomes (UN, 2006)

All of these realities, all of these uses, all of these dependencies mean that effective water management must include both a multitude and a combination of options and approaches; there is no silver bullet to cope with water-related challenges.

The global political context of the water debate

There are positive signs that water is taking centre stage internationally, and that political will is being built regionally and globally (e.g. EU, 2000) and globally (UN, 2002a, 2002b, 2003, 2006). Already in 1992, both the International Conference on Water and the Environment in Dublin, Ireland (United Nations Administrative Committee on Coordination and the Inter-Secretariat Group for Water Resources, 1992) and the United Nations Conference on Environment and Development (UNCED) in Rio de Janeiro, Brazil, emphasized that urgent action was needed on water. Chapter 18 of Agenda 21 stated:

> *The holistic management of freshwater as a finite and vulnerable resource, and the integration of sectoral water plans and programmes within a framework of national economic and social policy, are of paramount importance for actions in the 1990s and beyond.* (UN, 1992)

Even then, this was in many ways a reiteration of the statements already made at the first global water conference in Mar del Plata, Argentina, in 1977 (UN, 1977).

The United Nations Commission on Sustainable Development (CSD) maintained focus on freshwater and sanitation in the years following Rio (CSD, 1994, 1998a, 1998b). Former United Nations Secretary General Kofi Annan identified water as one of ten crucial issues to be addressed at the 2002 Johannesburg World Summit on Sustainable Development (WSSD) and, indeed, it was (UN, 2002a, 2002b). His successor, Ban Ki-moon, highlighted the need to focus more on water issues within the climate change agenda during the 2008 World Economic Forum, stressing that 'a shortage of water resources could spell increased conflicts in the future'. He added that 'governments must engage and lead, but the private sector also has a role to play'. The debate continues.

There are a range of political issues associated to water. Transboundary water management can either stir conflict or increase cooperation. Water as a human right continues to cause debates. Increasingly, access to a basic supply of safe drinking water is considered to be a universal human right (United Nations Committee on Economic, Social and Cultural Rights, 2003). Privatizations in the water sector are seen as the solution to some and a stigma to others. The fact that billions still lack access to water and sanitation can be seen as nothing less than a political scandal. Perhaps a reason could be that although water and sanitation targets are included in the Millennium Development Goals (MDGs) adopted in 2000 (sanitation was added at a later stage at the WSSD), the critical roles of water for achieving poverty, food, energy and health improvements are still severely overlooked.

Increasing pressure on water is a fact

Despite increased political attention, pressure on the world's water resources will increase and must be managed. The driving forces are population growth, more food and better diets, changing consumer preferences and disposable income, trade policies and socio-economic development, in general. The latter, although a driver, also influences society's ability to handle water scarcity and is perhaps the most critical factor for water resources management. Thus, the recent financial crisis will have an impact upon water resources management, although it is difficult to predict exactly how.

Extreme poverty affects approximately 1 billion people worldwide (UN, 2006) who, without reliable access to water, will remain poor. Supplying them and the additional 3 billion people on Earth by 2050 with safe drinking water and sanitation is a water resource, political and financial challenge. Beyond drinking water and sanitation, increased human needs and wants for food, energy and other goods and services will require even more water.

At the international conferences, many of the same messages have been repeated, and high sounding statements of concern have been given. New management approaches (e.g. integrated water resources management) have helped some regions to make progress in water supply and sanitation and enabled food production to keep pace with population growth (malnutrition is still more an issue of distribution); but, in general, progress on the ground has been slow.

This is partly because water issues are not well integrated in other policy and political processes guiding, for instance, trade, agricultural and energy subsidies, economic development, poverty eradication and health. Although the fundamental role that water plays in such sectors offers a potential opportunity (if the role is recognized, the resulting policy changes may affect water resources management positively and substantially), the potential will not be realized as long as planning and policy-making remains fragmented and sector oriented.

Water challenges for today and tomorrow

Scarcity is a complex concept

Is water scarcity/physical shortage a real problem? Although some statistics show that more than 200 million sub-Saharan Africans live in water-short countries (a figure predicted to increase to 700 million by 2025), the 2006 *World Water Development Report* shows low water stress and adequate supply for most of this region (UN, 2006) – at least on paper. Neither impression is wrong; each highlights how scarcity is not necessarily due to a physical lack of water. What is missing are investments and infrastructure, in particular, and the political will to take firm decisions, in general. Water can be scarce; but sensible decision-making for adapting to available resources and using them sustainably is even scarcer. The effects today are staggering; the implications for tomorrow are scary.

The scandal

On the individual human level, the water supply and sanitation challenges are most acute. As mentioned earlier, it is scandalous that almost 1 billion people lack access to safe drinking water and 2.4 billion lack adequate sanitation (WHO and UNICEF, 2008). Every year 12 million people die of easily preventable water-related diseases, including 3800 children who die every day (WHO, 2004). This human catastrophe caused by lack of clean water following poor infrastructure and inefficient water management also represents lost economic and social development opportunities.

To reach the Millennium Development Goal targets in water supply and sanitation requires about US$11.3 billion (WHO and UNICEF, 2005) annually. Yet, only one third of that is being invested. In addition to financial capital through aid as well as private investment, good and effective projects are prerequisites for building sustainable sanitary infrastructure, developing human capacity to deal with the issues locally, and promoting education about good hygiene, etc.

Integration is a must

The future of water resources management will not only have to take into account these gaps in drinking water and sanitation coverage. It will also have to deal with feeding and providing energy for 9 billion people; sustaining economic growth to alleviate poverty; and sustaining economic growth to increase living standards, all the while ensuring environmental sustainability. Issues cannot be solved one by one; integrated approaches in management, albeit easy to say and hard to apply, must be further developed.

There are roadmaps to a sustainable future in the form of top-notch, knowledge-based reports from which good policies and management strategies can be drawn. However, the prevalence of so many reports and analyses is also a weakness, for the planet is assessed through a sectoral or issue-oriented approach. It is fundamentally challenging for policy and decision-makers to find a comprehensive analysis (from the multitude of assessments on water, climate, biodiversity and energy) that presents both the interdependencies and conflicting interests.

Climate change – the ultimate challenge?

Climate change adaptation is a case and point. From 1992 to 2001, nearly 90 per cent of all natural disasters were of meteorological or hydrological origin (UN, 2006). However, our understanding of climate change impacts upon water resources is limited, in part because the interactions are complicated and because the effects are governed by a range of non-climate factors. Modified landscapes and infrastructure development as well as changes in hydrological systems (river modification) strongly influence the effects of climate variability and change. With increased flooding, for example, it is difficult for the planner to understand an essential question: how much of the increase is due to climate change and how much results form non-climate factors?

The most recent authoritative and comprehensive scientific assessment of climate change and the effects on water and society is presented in the *Fourth Assessment Report* of the Intergovernmental Panel on Climate Change (IPCC, 2007; and, in particular, Kundzewicz et al, 2007). The *Human Development Report* (HDR) 2007/2008 places climate change in a wider human development perspective (UNDP, 2007). In common, they state that our societies are at great risk and water is a critical factor. The IPCC (2007) predicts increased runoff (leading to more annual water availability) in the high latitudes of North America and Eurasia and in the tropics, while Mediterranean-style climates will see decreased runoff. Changes in the seasonality of runoff due to shifts in the snow/rain ratio in high latitudes and mountainous regions are expected with a high degree of certainty. The *Human Development Report* (UNDP, 2007), being more policy oriented, states more clearly that large areas of the world face imminent water stress and water availability for human settlements and agriculture will decrease.

Changes in precipitation are not the only important factor. Most mountain glaciers are retreating (Lemke et al, 2007) and, for now, this increases annual net flow of water in rivers. At the same time, however, their storage capacity decreases. When (or if) a glacier eventually disappears, the effects on the seasonal availability of water in downstream regions can be dramatic. The IPCC (2007) states that one sixth of the global population relies part of the year on meltwater from glaciers and permanent snow packs, so the implications are staggering.

For a water resources planner at the local or even national level, the IPCC predictions are still not very helpful: 'There is a scale mismatch between the large-scale climatic models and the catchment scale, which needs further resolution' (Kundzewicz et al, 2007). Although the IPCC stresses the uncertainties in its assessments of future changes in water resources due to climate change, as it should, it clearly points out that there are a range of challenges and risks that need to be considered in future adaptation strategies. These range from the particularly difficult situation related to basic water availability in semi-arid and arid regions, to the complex impacts upon ecosystems and the effects from climate change on the functioning and operation of infrastructure.

For the planner, non-climatic factors need to be considered as well. Adaptation strategies should not disregard such other factors or else the risk will be that investments are made in vain. Land-use changes, large-scale water diversions, changes in consumption patterns, changes in production (agriculture, industry), changes in population and population patterns, etc. will influence both water resources and local regional climates in various ways, sometimes even more than the local regional manifestation of the global climate change. Feedback systems, in reality, become an intrinsic web of relationships and outcomes. Understanding such links forms the foundation for any understanding of the water–climate–society interface.

The potential effects of climate change also need to be revived in a wider risk analysis setting. One in ten humans lives in coastal areas of less than 10m above sea level (McGranahan et al, 2007). These people are not now suddenly living at risk due to climate change. They have always been at risk in these areas; climate change simply changes the dynamic and adds an additional dimension. Clearly, efficient physical planning may be our most powerful adaptation tool.

Urbanization will also drive changes and often imply higher sensitivity to climate change. In 2007 humans for the first time became more urban than rural in real numbers, with more than 50 per cent of the world's population living in cities (UN, 2005), including 900 million in urban slums (UN, 2006). Urbanization adds complexity to understand climate change impact upon water by changing the physical properties (runoff, soil water and groundwater recharge, evaporation, etc.), but also by affecting water consumption trends and management opportunities. Both aspects need to be considered.

Food and energy

Maybe the key driver of our water future will be to cope with biomass production for both food and energy. The *Comprehensive Assessment* (Molden, 2007) called for drastic changes in how water is used and managed; otherwise there will not be enough water to meet the food, feed and fibre needs of humanity in the coming 50 years. Between 2000 and 2030 the production of food in developing countries needs to increase by 67 per cent. At the same time, a rise in water productivity should make it possible to limit the increase in agricultural water use to about 14 per cent (UN, 2006).

The recent rise of biofuels has altered the water agriculture equation dramatically, with far-reaching consequences unknown just a few years ago. The emergence of biofuels exemplifies the evolving way in which land and water resources are managed. It shows that the increasing complexity makes the task of accurately projecting future scenarios nearly impossible. Land previously used to grow food is now being used to grow crops for biofuel.

One thing this will do is affect the production and price of food crops and increase the amount of water used by agriculture (Molden, 2007). In areas experiencing water stress, biofuel production could reduce the availability of water for more basic needs of people and ecosystems. As Varghese (2007) states: 'the indiscriminate promotion of biofuel development as a "cheap and green" energy option may interfere with optimal water allocation, and/or the pursuit of appropriate public water policies that will help address the water crisis'. Although biofuel feedstock accounts for only 1 per cent of the total area under tillage, and a similar percentage of crop water use, production continues to grow rapidly.

The future: Hopeful or hopeless?

The solutions necessary to meet our current and future water-related challenges have been identified in a number of excellent assessments documented within this chapter. The challenge is now to combine this knowledge and translate it into good ideas for action and at the same time seek to resolve conflicting goals. An additional challenge is to reach out to those who may not have a direct interest in water issues, but whose decisions have major impacts upon them. The water sector must better consider the facts presented in assessments such as the *World Energy Outlook* (IEA, 2007), where a lack of access to a basic energy supply was presented as the obstacle to achievement of the MDGs. This outlook projects a different development path than do more directly water-oriented assessments.

There is enough water in the world for all human and ecosystem needs. Locally and regionally, the provision of enough water for drinking, sanitation, food and development can be difficult, particularly in rapidly growing urban areas in semi-arid regions. But looking behind the statistics it becomes clear that much can be done, quite quickly and even easily. In many cities in developing countries, for example, leakages in water systems (over 50 per cent) can be fixed. Water subsidies in agriculture (and energy for irrigation pumps) can be reduced as an incentive to make water use more efficient. Some estimates show that up to 40 per cent of all the food being produced never reaches the consumer or is thrown away before being eaten (Lundqvist et al, 2008). Both producer and consumer efficiency can be improved.

But we also have to look at new perspectives and ideas realistically and carefully. One interesting example is illustrated by De La Torre and He (2007). If all oil-based transportation fuels should be replaced by fuels derived from biomass, about 30 million barrels of ethanol and 23 million barrels of biodiesel would be required per day. For such production, 300 million hectares of sugarcane (assuming yields similar to that in Brazil) and 590 million hectares of corn (assuming yields similar to that in the US) need to be planted to meet the ethanol needs alone. To put such numbers in a comprehensive context, this equals about 15 times the current world planting of sugarcane and 5 times the current corn planting. To meet the biodiesel demand, 225 million hectares of palm, or 20 times the current area, need to be planted. What would be the water demands? Not yet calculated! Of course, this would not be a realistic development; but it illustrates how important it is to include land and water aspects into the energy and climate change mitigation discussions.

So what can we do? There are many options but no blueprint solutions. Future investments in land and water systems will have to be much more flexible and responsive to climate opportunities. Investments should be assessed 'beyond the sector', as there may be competition (the water–food–energy nexus) and strategic opportunities. If carefully implemented, investments in, for example, climate change adaptation could represent 'no regrets' options, despite uncertainty, as they provide immediate benefits while at the same time making society better prepared for and less vulnerable to future changes, including climate change

We also need to realize that access to information relevant for policy and management is a strategic issue. Data and state-of-the-art knowledge needs to be more efficiently translated into policy- and management-relevant information. The issue of scale will be fundamentally important and knowledge transfer and capacity-building need to increasingly focus on the user's level.

The multilateral system also needs to be strengthened. Increasingly, challenges have global dimensions (e.g. the food crisis, climate change, the financial crisis, the energy crisis) and require increasing international cooperation. Multilateral systems also need to be supported to increase collaboration among different actors, which are specialized in specific areas. Water resources issues are cross-cutting and need to be addressed from that perspective. United Nations Water represents an innovative and flexible mechanism to promote such collaboration within the UN system and with key partners.

Is the lack of financial resources a real problem? Maybe this question is a bit provocative in this day and time with a global financial crisis crippling economic development. Although comparisons between different sectors can be unfair to make, the Stockholm International Peace Research Institute (SIPRI, 2007) calculates that world military expenditures in 2006 reached US$1204 billion and represents a 3.5 per cent increase in real terms since 2005 and a 37 per cent increase over the last ten years. The average spending per capita is US$184. Again, reaching the MDG targets on water would cost US$11.3 billion annually. SIPRI also notes that the ratio of military spending to social spending was highest in those countries with the lowest per capita incomes. Such numbers show that there are financial resources available, if only the priorities would change. Neither water nor resources are in short supply, albeit unevenly distributed. Political will to promote change is still lacking – an overused generalization, but still true. To make change happen, good arguments and relevant decision-making support are needed, as is a considerable dose of optimism.

Note

1 Evaporation and transpiration combined is often referred to as evapotranspiration.

References

Ban Ki Moon (2008) World Economic Forum www.un.org/apps/news/story.asp?NewsID=25398&Cr=davos&Cr1

Baresel, C. and G. Destouni (2005) 'Novel quantification of coupled natural and cross sectoral water and nutrient/pollutant flows for environmental management', *Environmental Science and Technology*, vol 39, pp6182–6190

CSD (Commission on Sustainable Development) (1994) *Report of the Commission on Sustainable Development on Its Second Session*, 16–27 May 1994, E/CN.17/1994/20, New York, NY

CSD (1998a) *Report on the Intersessional Ad Hoc Working Group on Strategic Approaches to Freshwater Management*, 23–27 February 1998, E/CN.17/1998/13, New York, NY

CSD (1998b) *Report on the Sixth Session*, 22 December 1997 and 20 April–1 May 1998, Economic and Social Council Official Records, 1998, Supplement No 9, E/CN.17/1998/20

Cosgrove, W. J. and F. R. Rijsberman (2000) *World Water Vision: Making Water Everybody's Business*, World Water Council, Earthscan, London

De La Torre, U. D. and L. He (2007) 'Is the expansion of biofuels at odds with the food security of developing countries?', *Biofuels, Bioproducts and Biorefining*, vol 1, pp92–102, www.interscience.wiley.com

EU (European Union) (2000) *EU Water Framework Directive: Directive 2000/60/EC of the European Parliament and of the Council Establishing a Framework for the Community Action in the Field of Water Policy*, EU Official Journal (OJ L 327, 22 December 2000)

FAO (United Nations Food and Agriculture Organization) (2006) *Irrigation Data Collection*, FAOSTAT Statistics Database, www.fao.org

Gleick, P. H. (2003) *Water in Crisis*, Oxford University Press, Oxford, UK

IEA (International Energy Agency) (2007) *World Energy Outlook 2007*, OECD and IEA, Paris, France

IPCC (Intergovernmental Panel on Climate Change) (2007) *IPCC Fourth Assessment Report*, UN, Geneva

Koplin, D. W., E. T. Furlong, M. T. Meyer, E. M. Thurman, S. D. Zaugg, L. B. Barber and H. T Buxton (2002) 'Pharmaceuticals, hormones, and other organic wastewater contaminants in US streams, 1999–2000: A national reconnaissance', *Environmental Science and Technology*, vol 36, pp1202–1211

Kundzewicz, Z. W., L. J. Mata, N. W. Arnell, P. Döll, P. Kabat, B. Jiménez, K. A. Miller, T. Oki, Z. Sen and I. A. Shiklomanov (2007) 'Freshwater resources and their management', in L. Parry, O. F. Canziani, J. P. Palutikof, P. J. van der Linden and C. E. *Climate Change 2007: Impacts, Adaptation and Vulnerability, Contribution of Working Group II to the Fourth Assessment Report of the Intergovernmental Panel on Climate Change*, Cambridge University Press, Cambridge, UK, pp173–210

Lemke, P., J. Ren, R. Alley, I. Allison, J. Carrasco, G. Flato, Y. Fujii, G. Kaser, P. Mote, R. Thomas and T. Zhang (2007) 'Observations: Changes in snow, ice and frozen ground', in S. Solomon, D. Qin, M. Manning, Z. Chen, M. Marquis, K. B. Averyt, M. Tignor and H. L. Miller (eds) *Climate Change 2007: The Physical Science Basis. Contribution of Working Group I to the Fourth Assessment Report of the Intergovernmental Panel on Climate Change*, Cambridge University Press, Cambridge, UK, pp337–384

Lindgren, G. A., G. Destouni and A. Darracq (2007) 'The inland subsurface water system role for coastal nitrogen load dynamics and abatement responses', *Environmental Science and Technology*, vol 41, no 7, pp2159–2164

Lundqvist, J., C. de Fraiture and D. Molden (2008) *Saving Water: From Field to Fork-Curbing Losses and Wastage in the Food Chain*, Stockholm International Water Institute Policy Brief, Stockholm

McGranahan, G., D. Balk and B. Anderson (2007) 'The rising tide: Assessing the risks of climate change and human settlements in low elevation coastal zones', *Environment and Urbanization*, vol 19, no 1, pp17–37

MEA (Millennium Ecosystem Assessment) (2005a) *Ecosystems and Human Wellbeing: Synthesis*, Island Press, Washington, DC

MEA (2005b) *Ecosystems and Human Wellbeing: Current State and Trends, Volume 1*, Island Press, Washington, DC

Molden D (2007) *Water for Food, Water for Life: A Comprehensive Assessment of Water Management in Agriculture*, Earthscan, London, and International Water Management Institute, Colombo
Shiklomanov, I. A. (2000) 'Appraisal and assessment of world water resources', *Water International*, vol 25, pp11–32
SIPRI (Stockholm International Peace Research Institute) (2007) *The SIPRI Yearbook: Armaments, Disarmaments and International Security*, SIPRI, Stockholm
Smakhtin, V., C. Revenga and P. Döll (2004) *Taking into Account Environmental Water Requirements in Global Scale Water Resources Assessments: Comprehensive Assessment of Water Management in Agriculture*, International Water Management Institute, Colombo
UN (United Nations) (1977) *Report of the United Nations Water Conference*, UN, Mar del Plata, 14–25 March 1977, E/CONF. 70/29
UN (1992) *Earth Summit: Agenda 21. The United Nations Programme of Action from Rio*, Chapter 18 'Protection of the quality and supply of freshwater resources: Application of integrated approaches to the development, management and use of water resources', pp166–185, www.un.org/esa/sustdev/agreed.htm
UN (2002a) *Report of the World Summit on Sustainable Development Johannesburg, South Africa*, 26 August–4 September 2002, UN, New York, NY
UN (2002b) *Johannesburg Plan of Implementation of the World Summit on Sustainable Development*, UN, New York, NY
UN (2003) *World Water Development Report: Water for Life, Water for People*, UNESCO, Paris, and Berghahn Books, Barcelona, Spain
UN (2005) *World Urbanization Prospects: The 2005 Revision*, Population Division, UN Department of Economic and Social Affairs, New York, NY
UN (2006) *World Water Development Report 2: Water, a Shared Responsibility*, UNESCO, Paris
United Nations Administrative Committee on Coordination and the Inter-Secretariat Group for Water Resources (1992) *The Dublin Statement and Report of the Conference*, International Conference on Water and the Environment: Development Issues for the 21st century, 26–31 January 1992, Dublin, Ireland
UN Committee on Economic Social and Cultural Rights (2003) *General Comment No 15 (2002): The Right to Water*, E/C.12/2002/11, United Nations Social and Economic Council, Geneva
UNDP (United Nations Development Programme) (2007) *Human Development Report 2007/2008 Fighting Climate Change*, UN, New York, NY
Varghese, S. (2007) *Biofuels and Global Water Challenges*, Institute for Agriculture and Trade Policy, Minnesota
WHO (World Health Organization) (2004) *Water, Sanitation and Hygiene Links to Health*: Facts and Figures, WHO, Geneva
WHO and UNICEF (2005) *Water for Life: Making It Happen*, WHO/UNICEF Joint Monitoring Programme for Water Supply and Sanitation, Geneva
WHO and UNICEF (2008) *Progress on Drinking Water and Sanitation, Special Focus on Sanitation*, WHO/UNICEF Joint Monitoring Programme for Water Supply and Sanitation, Geneva

11
Handling or not Handling the Food Crisis

Jaques Diouf[1]

In 2008 we had before us a world food crisis that had tragic social and political consequences in different continents, with riots and deaths that endangered world peace and security. Those sad events are, however, but the chronicle of disaster foretold. In 1996, at the Food Summit, 112 heads of state and government and the representatives of 186 members of the United Nations Food and Agriculture Organization (FAO) solemnly pledged to reduce by half the number of hungry in the world by the year 2015 and adopted a programme to achieve that target. But already in 2002, we had to convene a second world summit to draw the international community's attention to the fact that resources to finance agricultural programmes in developing countries were decreasing, instead of rising. With such a trend, the summit target would not be reached in 2015, but in 2150. An Anti-Hunger Programme with financial requirements estimated at US$24 billion per year had been prepared for that meeting. Global food production must double to feed a world population currently standing at 6 billion and expected to rise to 9 billion by 2050.

Today, the facts speak for themselves: from 1980 to 2005, aid to agriculture fell from US$8 billion (2004 basis) in 1984 to US$3.4 billion in 2004, representing a reduction in real terms of 58 per cent. Agriculture's share of official development assistance (ODA) fell from 17 per cent in 1980 to 3 per cent in 2006. The international and regional financial institutions saw a drastic reduction in resources allocated to the activity that constitutes the principal livelihood of 70 per cent of the world's poor. In one telling case, the loan portfolio to agriculture of one institution plummeted from 33 per cent in 1979 to 1 per cent in 2007. In cooperation with the FAO, the developing countries did, in fact, prepare policies, strategies and programmes that, if they had received appropriate funding, would have ensured world food security.

Following a meeting of African experts in December 2001 in Rome, the ministers for agriculture met at the FAO Regional Conference for Africa in Cairo in February 2002, and again in Maputo just before the July 2003 African Union Summit. On that occasion, the heads of state and government adopted the Comprehensive Africa Agriculture Development Programme (CAADP) and its companion documents prepared with the support of the FAO (African Union, 2003). The programme requires an investment of US$25 billion per year for water control, rural infrastructure, trade capacity, increased crop production and reduced hunger, agricultural research and the dissemination of technology, animal production, forestry, fisheries and aquaculture. In this context, 51 African countries, with the support of the FAO, prepared national medium-term investment programmes (NMTIPs) and bankable investment project profiles (BIPPs).

The regional economic communities – the Union Economique et Monétaire Ouest Africaine (UEMOA), the Economic Community of West African States (ECOWAS), the Southern African Development Community (SADC), the Common Market for Eastern and Southern Africa (COMESA), the Intergovernmental Authority on Development (IGAD) and the Arab Maghreb Union (AMU) – have, with the FAO's support, also prepared regional food security programmes that focus on intra-regional trade and the World Trade Organization (WTO) sanitary and phytosanitary standards, based on the rules established by the World Health Organization (WHO) and FAO for consumer protection in the framework of the Codex Alimentarius[2] and the International Plant Protection Convention (FAO and WHO, 1963).

Following implementation of the pilot phases of national and regional food security programmes in the countries of the Caribbean Community and Common Market (CARICOM), and Central and South America, the Ibero-American Summit approved, in November 2006, in Montevideo, Uruguay, the Initiative called Hunger-Free Latin America and the Caribbean by 2025 (FAO, 2006). Similar regional programmes were prepared, in cooperation with the FAO, in Central Europe and Central Asia for the Organization of the Black Sea Economic Cooperation and the Economic Cooperation Organization.

Plans, programmes and projects – well and good – thus do exist to address food security, even though they may require further refinement and updating. But sadly, the international community only reacts when the media beams the painful spectacle of world suffering into the homes of the wealthy countries.

Based on world agricultural statistics and the projections that the FAO is responsible for preparing, already in September 2007, I alerted public opinion to the risks of social and political unrest due to hunger. On 17 December 2007, to avoid jeopardizing the 2008 agricultural season, I launched an appeal for the mobilization of US$1.7 billion in grants to enable the farmers of poor countries to have access to the fertilizer, seeds and animal feed that had risen in price by 98, 72 and 60 per cent, respectively. All of this was in vain. Despite broad press coverage, and correspondence to the member states and financial institutions, only a few countries, such as Spain, offered their immediate support to agricultural production. I would like to pay tribute to those countries which did assist us.

It was only when the destitute and those excluded from the banquets of the rich took to the streets to voice their discontent and despair that the first reactions in support of food aid began to emerge.

The causes and consequences of the present crisis have been explained at length, so I shall not return to them. What is important today is to realize that the time for talking is long past. It is time for action.

The United Nations Secretary General set up and chairs the Task Force of the United Nations system, the Bretton Woods institutions and other international organizations to bring a coordinated response to the food crisis (UN, 2008), of which I am the vice chair. The Comprehensive Framework for Action prepared by the task force provides guidelines on the needs that will be specified, country by country, with the assistance of the local representatives of the FAO, the World Food Programme (WFP), the International Fund for Agricultural Development (IFAD) and the World Bank, in cooperation with the governments. In connection with this, on 29 April 2008 in Berne, the Secretary General of the United Nations presented to the press the communiqué approved by the United Nations Chief Executives Board (CEB) for coordination on the immediate need to mobilize the necessary resources and deal with the food crisis. Of course, there was a pressing need, despite escalating prices, to maintain the volume of food aid for 88 million people. We thank the countries that contributed so generously to meet the required US$755 million in this regard.

But there are 862 million people in the world who do not have adequate access to food. They need to enhance their living conditions in dignity, working with the resources of today's generation. They need high-yield seeds, fertilizer, animal feed and other modern inputs. They cannot continue to toil as in the Middle Ages, under conditions of uncertainty and exposure to the whims of the weather. Investment is therefore needed in rural infrastructure, especially for water control with irrigation and drainage, considering, for example, that 96 per cent of arable land in sub-Saharan Africa depends upon rainfall. They need storage facilities to avoid harvest losses that can amount to 40 to 60 per cent for certain crops. Rural roads are essential to bring in modern factors of production and enable harvests to reach domestic and regional markets at competitive prices.

The current food crisis goes beyond the traditional humanitarian dimension, which has an eminently ethical foundation. This time it also affects the developed countries. Rising inflation is 40 to 50 per cent the result of higher food prices. In a context of high and accelerated growth of gross domestic product (GDP) of the emerging countries, we must seek sustainable and viable global solutions that will narrow the gap between global food supply and demand. If we do not urgently take the courageous decisions that are required in the present circumstances, the restrictive measures taken by producer countries to meet the needs of their populations, the impact of climate change and speculation on future markets will place the world in a dangerous situation. Whatever the extent of their financial reserves, some countries might not find food to buy.

The structural solution to the problem of food security in the world lies in increasing production and productivity in the low-income food-deficit countries. This calls for innovative and imaginative solutions, besides official development assistance. Partnership agreements are needed between countries that have financial resources, management capabilities and technologies, and countries that have land, water and human resources. Only in this way will it be possible to ensure balanced international relationships for sustainable agricultural development.

The challenges of climate change, bioenergy, transboundary animal and plant diseases and agricultural commodity prices can only be met through frank dialogue based on objective analysis devoid of partisan and short-term interests.

Yet, obligation to truth compels me to note certain facts:

- Nobody understands how a carbon market of US$64 billion can be created in the developed countries to offset global warming, while no funds can be found to prevent the annual deforestation of 13 million hectares, especially in the developing countries whose tropical forest ecosystems act as carbon sinks for some 190 gigatonnes of CO_2.
- Nobody understands how US$11 billion to $12 billion in subsidies in 2006 and protective tariff policies have had the effect of diverting 100 million tonnes of cereals from human consumption, mostly to satisfy a thirst for fuel for vehicles.
- Nobody understands how in a time of globalization of trade, with the notable exception of avian influenza that could lead us to human calamity, there has been no significant investment in the prevention of Newcastle disease; foot and mouth disease; Rift Valley fever; contagious bovine pleuropneumonia; the pests of small ruminants; bluetongue disease; African swine fever; tropical bont tick and the new world screwworm; but also wheat stem rust, which since 1999 has spread from Uganda to Iran and could reach India, Pakistan and China; the fruit fly; and finally desert locusts, a scourge familiar since the time of the Pharaohs.
- But, above all, nobody understands how, first, the Organisation for Economic Co-operation and Development (OECD) countries have created a distortion of world markets with the US$372 billion spent in 2006 on supporting their agriculture; second, that in a single country food wastage can amount to US$100 billion annually; third, that the excess consumption by the world's obese costs US$20 billion annually, to which must be added indirect costs of US$100 billion resulting from premature death and related diseases; and, finally, that in 2006 the world spent US$1200 billion on the purchase of arms.

Against this backdrop, how can we explain to people of good sense and good faith that it is not possible to find US$30 billion a year to enable 862 million hungry people to enjoy the most fundamental of human rights: the right to food and, thus, the right to life? It is resources of this order of magnitude that would make it possible definitively to lay to rest the spectre of conflicts over food that are looming on the horizon.

The problem of food insecurity is a political one. It is a question of priorities in the face of the most fundamental of human needs. It is the choices made by governments that determine the allocation of resources.

Notes

1 This chapter is based on a speech delivered at the High-Level Conference on World Food Security, Rome, 2008.

2 The Alimentarius Commission was created in 1963 by the FAO and WHO to develop food standards and guidelines in order to protect the health of consumers and ensure fair trade practice.

References

African Union (2003) *Comprehensive Africa Agriculture Development Programme (CAADP)*, African Union, Mauritius

FAO (United Nations Food and Agriculture Organization) and WHO (World Health Organization) (1963) *Codex Alimentarius and the International Plant Protection Convention*, www.codexalimentarius.net/web/index_en.jsp

FAO (2006) *Hunger-Free Latin America and the Caribbean by 2025 – Caracas, Venezuela*, FAO, Rome

UN (United Nations) (2008) *Secretary General's Task Force on the Global Food Security Crisis*, www.un.org/issues/food/taskforce/

12
Climate Change and Biodiversity Loss: The Great New Threats to Peace, Security and Development

Ahmed Djoghlaf

Since the Industrial Revolution, humanity has entered an era of global changes that is transforming the face of the Earth. The effects of that revolution will ripple down for centuries. Rising living standards are taking their toll on the natural resource base that underpins them. As a result, humanity's footprint on the environment has never been as significant as it is now. The pressures on the planet's natural functions caused by human activity have reached such levels that ecosystems' ability to satisfy the needs of future generations (a necessary component of achieving sustainable development; WCED, 1987) is seriously, and perhaps irreparably, compromised. Humanity is facing two main global challenges that are interconnected: climate change and biodiversity loss. These challenges have further implications for peace, security and development (WBGU, 2007).

There is now general agreement that human activities are affecting the planet's climate. In turn, climate change is affecting biodiversity and people around the world. Observational evidence from all continents and most oceans shows that many natural systems are being affected by climate change, particularly temperature increases and changes in precipitation. Recent warming is strongly affecting biological systems, including changes such as earlier timing of spring events, leaf unfolding, bird migration and egg laying, as well as pole-ward and upward shifts in ranges in plant and animal species.

The *Fourth Assessment Report* of the Intergovernmental Panel on Climate Change (IPCC, 2007) highlights a number of regional impacts that could have severe repercussions upon the environment and people. For example, by 2020, between 75 and 250 million people in Africa are projected to be exposed to increased water stress due to climate change (UNESCO, 2006). By 2050, climate change will lead to gradual replacement of tropical forest by savannah in eastern Amazonia. Semi-arid vegetation will tend to be replaced by arid land vegetation. There is a risk of significant biodiversity loss in many areas of tropical Latin America. In North America, coastal communities and habitats will be increasingly stressed by climate change impacts interacting with development and pollution.

People in the small island developing states (SIDS) areas are observing declines in reef health due to climate change, which are adversely affecting traditional fisheries and local livelihoods. In the Arctic, approximately 400,000 indigenous people are facing the impacts of a warming at almost twice the rate of the global average. The Sami people, for example, are already observing changes in species composition within reindeer pasture lands, while the Inuit are altering their hunting practices in response to changing habits and habitats among their prey. The impacts of climate change are also of particular concern to people living in dry and sub-humid lands, ranging from the deserts and savannahs of Africa to the Arctic tundra. In fact, changing precipitation regimes, increased heat stress, and the increased frequency and intensity of extreme weather events are contributing to desertification and the associated loss of ecosystem productivity throughout the world's dry and sub-humid lands.

The Stern Review, produced by a team of economists led by the former World Bank Chief Economist Lord Nicholas Stern, has said that if rapid actions are not taken to mitigate climate change, there could be significant costs resulting from worldwide recession, mass migration and social upheaval (Stern et al, 2006). Indeed, climate change impacts upon natural resources may trigger conflicts over the distribution of resources, especially water and land. Approximately 1.1 billion people do not have access to adequate quantities of safe freshwater resources. This situation could worsen in some regions of the world as climate change is likely to alter the precipitation patterns and reduce the quantity of water available.

The demand for water for agriculture in a changing climate could also worsen the problems associated with desertification. To satisfy the growing demand for food, production of food crops in developing countries is projected to increase by 67 per cent by 2030. Irrigated agriculture in these countries would consequently account for over 70 per cent of the projected increase in cereal production. Taking into account increased cropping density and rise in productivity, the amount of freshwater to be appropriated for irrigation is expected to increase by about 14 per cent in 2030 (UNESCO, 2006).

With increasing demand for food, large-scale degradation of water resources represents a serious threat to ecosystems. Over the past 50 years, technological advances in agriculture have led to increased crop yields. Nevertheless, agricultural productivity is still highly dependent upon climate as rainfall, temperature and solar radiation condition plant growth. Climate change is thus projected to affect regional food security, especially in developing countries. Changes in rainfall patterns and extreme weather events are likely to diminish crop yields in many areas.

At the same time, rising sea levels, causing loss of coastal land and saline water intrusion, can also reduce agricultural productivity. Coral bleaching and increased calcification of coral are likely to reduce fisheries, further threatening food security. Changes in habitat are already negatively affecting the supply of protein available from hunting, especially in the Arctic region.

Invasive species are another important threat to the food supply and livelihoods. As a result of climate change, many ecosystems will be more vulnerable to invasive species, pests and diseases. The extent to which invasive pests and diseases either reduce the domestic supply of food directly or restrict a country's international trade could harm its food security, especially for developing countries. Across Africa, for instance, the invasive alien species of the genus *Striga*, a parasitic weed that impedes maize growth by attacking the roots, costs between US$7 million and US$13 million annually and affects more than 100 million people (UNEP, 2006).

In addition to the issues linked to food security, climate change could also increase health hazards. Through geographic changes in weather patterns, rainfall and temperature, climate change is predicted to dramatically increase the extent and prevalence of some vector-borne diseases, such as malaria and dengue fever. Extreme weather events may also increase vulnerability to water, food or human-borne diseases such as cholera and dysentery, and lead to increases in heat-related mortality and illness. Floods and droughts may result in declining quantity and quality of drinking water. Clean and adequate quantities of water are a prerequisite for good health and prevention of child mortality; without this basic necessity, morbidity and mortality rates rise. Climate change may exacerbate malnutrition by reducing natural resource productivity.

According to the IPCC, climate change could lead, towards the end of the present century, to an increase in global average sea level of up to 59cm (Parry et al, 2007). The very existence of several island countries such as Tuvalu or Kiribati is under threat. In November 2001, Tuvalu was the first country to make arrangements for the evacuation of its population because of rising water levels. Looking at future predictions of sea-level rise, Tuvalu may, sadly, not be the last. Therefore, because of climate change, a number of sovereign nations will eventually vanish from the surface of the Earth. No technology, whatever its level of sophistication and advancement, will be able to prevent this.

Climate change and biodiversity loss represent a serious threat to humanity, particularly to people living in developing countries. The countries that contribute the least to global warming will suffer the most. The countries that are already struggling to feed and provide shelter to their populations will have to do with even more limited resources. In this way, climate change and biodiversity loss represent a great impediment to the achievement of the Millennium Development Goals (MDGs) (UN Convention on Biological Diversity, 2007a).

Furthermore, climate change can be a factor in civil, and international, conflict. Environmental stress is seldom the only cause of major conflicts within or among nations. Nevertheless, they can arise from the marginalization of sectors of the population and from ensuing violence. This occurs when political processes are unable to handle the effects of environmental stress resulting, for example, from climate change. Environmental stress can thus be an important part of the web of causality associated with any conflict and in some cases can act as a catalyst.

The issue of environmental degradation brings new light to bear on the traditional concepts of international peace and security. Global changes, such as climate change and biodiversity loss, call for alliances. That is why, since 1989 following the preparations for the Earth Summit, there have been major calls for environmental issues or environmental degradation to be included on the agenda of the United Nations Security Council. On 17 April 2007, the United Nations Security Council for the first time addressed the question of climate change in order to build a shared understanding of the relationship between energy, security and climate.

The Convention on Biological Diversity (CBD), the first and most important global agreement on the conservation and sustainable use of biodiversity, links traditional conservation efforts to the development goal of using biological resources in a sustainable manner. Fifteen years have passed since its adoption. Even though significant progress has been achieved at the local level, the relationship between the environment and development still has not been globally reconciled and important aspects of our societies such as health, peace, security and development are more than ever under threat.

At the initiative of former Secretary General of the United Nations Kofi Annan, the 2010 biodiversity target of achieving a significant reduction in the rate of loss of biodiversity by that date was incorporated within the Millennium Development Goals as a new target under Goal 7 ('Ensure environmental sustainability'). Specifically, the conservation of biodiversity is recognized for its positive contribution to poverty alleviation and development. The Conference of the Parties to the CBD (UN Convention on Biological Diversity, 2007b), at its eighth meeting, highlighted the importance of integrating climate change considerations within all relevant national policies, programmes and plans, and to rapidly develop tools for the implementation of biodiversity conservation activities that contribute to climate change adaptation.

Indeed, the resilience of ecosystems to climate change can be enhanced and the risk of damage to human and natural ecosystems reduced through the adoption of biodiversity-based adaptation and mitigation strategies. Examples of such strategies include the maintenance and restoration of native ecosystems, the protection of ecosystem services, the management of habitats for endangered species, the creation of refuges and buffer zones, and the establishment of networks of terrestrial, freshwater and marine protected areas that take into account projected changes in climate.

In order to complement the steps already taken to link biodiversity and climate change, the Conference of the Parties to the CBD adopted decisions IX/16A–D on biodiversity and climate change (UN Convention on Biological Diversity, 2008). The decisions recognize that actions at the national level are best able to build synergies between biodiversity, climate change and land degradation and propose an indicative list of activities that parties can implement to this end. To support such activities, the Conference of the Parties also calls for collaboration between the secretariats of the three Rio conventions on communication and educational activities.

The decisions also outline a number of areas where additional research is required, including the impacts of multiple nutrient loading upon biodiversity, the impacts of climate change upon plant pests, the role of traditional and local knowledge in adapting to climate change, and the links between biodiversity and reducing emissions from deforestation and forest degradation.

The decisions urge support for capacity-building activities. These include raising public awareness in order to enable developing countries to implement activities related to the impacts of climate change, as well as publicizing the positive and negative impacts of climate change mitigation and adaptation activities upon biodiversity.

Finally, the decision establishes an Ad Hoc Technical Expert Group on Biodiversity and Climate Change. The group has a mandate to develop scientific and technical advice on biodiversity in so far as it relates to climate change and decision 1/CP.13 of the Conference of the Parties to the United Nations Framework Convention on Climate Change (UNFCCC) (the Bali Action Plan), as well as its Nairobi work programme on impacts, vulnerability and adaptation to climate change. Working within the context of the Bali Action Plan and the Nairobi work programme supports the enhanced implementation of synergies.

The implementation of these and other activities relating to climate change and biodiversity will be evaluated at the tenth meeting of the Conference of the Parties to the CBD, to be held in 2010, thus providing another opportunity to examine biodiversity and climate change from an international policy perspective.

The need for collaboration and scientific partnerships is crucial to address the global challenge of climate change and biodiversity loss. In March 2007, roundtable discussions on the further strengthening of the interlinkages between biodiversity and climate change were held in Montreal. For the first time members and past chairs of the bureau of the CBD Subsidiary Body on Scientific, Technical and Technological Advice (SBSTTA) and experts from the IPCC met to discuss commonalities and opportunities for enhanced cooperation between the two bodies.

As governments around the world develop their climate adaptation strategies, failure to appreciate the importance of biodiversity can potentially lead to counterproductive measures.

The link between climate change and biodiversity is one of survival for humankind. Under the CBD, virtually all governments have agreed to the very challenging target of slowing the loss of biodiversity by 2010. The time has come to recognize that meeting this target is an essential step in allowing future generations to prosper in the less stable climate they will unfortunately inherit. The need for action in order to achieve a more sustainable future has never been so urgent. There is no time for delay.

Irrespective of the future approach adopted by the United Nations for addressing the issue of climate change and biodiversity loss, it is clear that the security dimension of this issue cannot be ignored. Due to its magnitude and its impact, climate change seems to fall more and more within the context of Chapter VII of the United Nations Charter relating to acts of aggression against world peace. The designation of Wangari Maathai as the winner of the Nobel Peace Prize in 2004 was a step in the right direction. She is the first environmental activist to win the award and the first African woman to receive any Nobel Prize. When the Nobel Peace Prize was presented to Al Gore, the former vice president of the US, and to the IPCC, the prestigious Norwegian Nobel Committee recognized, for the second time in its history, the environmental dimension of the concept of peace and security. As Professor Ole Danbolt Mjos, chair of the Norwegian Nobel Committee, highlights: 'environment protection has become yet another path to peace'.

References

IPCC (Intergovernmental Panel on Climate Change) (2007) *IPCC Fourth Assessment Report*, UN, Geneva

Mjos, O. D. (2004) Presentation speech of the Nobel Peace Prize, Oslo

Parry, C., M. Canziani, O. Palutikof, J. van der Linden and P. Hanson (eds) (2007) *Climate Change 2007: Impacts, Adaptation and Vulnerability: Contribution of Working Group II to the Fourth Assessment Report of the Intergovernmental Panel on Climate Change*, Cambridge University Press, Cambridge, UK

Stern, N., S. Peters, V. Bakhshi, A. Bowen, C. Cameron, S. Catovsky, D. Crane, S. Cruickshank, S. Dietz, N. Edmonson, S.-L. Garbett, L. Hamid, G. Hoffman, D. Ingram, B. Jones, N. Patmore, H. Radcliffe, R. Sathiyarajah, M. Stock, C. Taylor, T. Vernon, H. Wanjie and D. Zenghelis (2006) *Stern Review: The Economics of Climate Change*, HM Treasury, London

UN Convention on Biological Diversity (2007a) *International Day for Biological Diversity: Biodiversity and Climate Change*, UN, Montreal

UN Convention on Biological Diversity (2007b) *Biodiversity, Climate Change and the Millennium Development Goals (MDGs). Factsheet: Biodiversity and Climate Change: Links with Health*, UN, Montreal

UN Convention on Biological Diversity (2008) *Biodiversity and Climate Change*, Decision IX/16, UN, Montreal

UNEP (United Nations Environment Programme) (2006) *Africa Environment Outlook 2. Factsheet on Invasive Alien Species*, www.unep.org/dewa/africa/aeo2_launch/index.asp

UNESCO (United Nations Educational, Scientific and Cultural Organization) (2006) *The United Nations World Water Development II. Water: A Shared Responsibility*, UNESCO and Berghanhn Books, Paris

WBGU (German Advisory Council on Global Change) (2007) *New Report: Climate Change as a Security Risk. Climate Change Threatens International Stability and Security*, Press release, WBGU, Berlin

WCED (World Commission on Environment and Development) (1987) *Our Common Future: Report of the World Commission on Environment and Development – the Brundtland Report*, Chapter 11 'Peace, Security, Development, and the Environment', Oxford University Press, Oxford, UK

13
Climate Change: A New Threat to Middle East Security

Gidon Bromberg, Ladeene Freimuth, Nader al Khateeb and Munqeth Mehyar

Introduction

With the Middle East being the world's most water-stressed region, the projected impacts of climate change, such as more extreme weather events, decreased precipitation and sea-level rise, will contribute to even greater water stress in the region, with severe environmental, economic, political and security implications.

That is, climate change is likely to act as a 'threat multiplier', exacerbating water scarcity and, as a result, tensions over water between nations linked by hydrological resources, geography and shared political boundaries (CNA Corporation, 2007). 'In fact, the enormously intricate water politics of the [Middle East] region have been aptly described as a hydro-political security complex' (Campbell et al, 2007).

The following factors will play a role in determining the likelihood for greater conflict or cooperation in this region, which already possesses some of the greatest political tensions in the world, as climate impacts become more significant:

- the existence of water agreements, and their degree of *sustainability*, including the ability of parties to deal with extreme circumstances, such as longer periods of drought;
- the influence of destabilizing economic and political factors (e.g. unemployment and mass migration due to agricultural decline and the large-scale flooding of agricultural areas);
- the extent of national economic and political development, including the degree to which local institutional structures and infrastructure exist;

- a given political entity's ability to mitigate and/or adapt to climate change;
- power relationships between the parties involved; and
- whether it is politically expedient at a given time to cooperate (or to continue to cooperate) over water resources.

Keeping these factors in mind, climate change provides both challenges and opportunities for cross-border cooperation to ameliorate and prevent the problems that are already occurring and are projected to intensify. For this reason, water issues, for example, have been an important part of all peace talks in the region. The following are some highlights of the unique aspects, as well as some of the gaps, in the water-sharing agreements that do exist, and some of the anticipated problems where such agreements do not yet exist, as the impacts of climate are expected to intensify.

Water-sharing agreements

The Middle Eastern governments believe that a lack of water will constrain their opportunities for development and, thus, endanger domestic political stability as well as relations with their neighbours. This belief has meant that efforts have been under way since the early 1950s to achieve agreements over water, despite larger ongoing political tensions or conflicts (Brauch, 2003, p729).

Israel–Jordan Peace Treaty and Water-Sharing Agreement

Jordan and Israel signed a peace treaty in 1994 that contains a water-sharing provision that aims to achieve a 'comprehensive and lasting settlement of all the water problems' between the two countries through mutual recognition of their 'rightful allocations' to water from the Jordan River and the Yarmouk River (Israel–Jordan Peace Treaty, Article 6, 1994b). It also aims to achieve mutual cooperation in the development of existing and additional water resources. Specifically, it allocates:

- for Israel – 25 million cubic metres of water per year from the Yarmouk River (Israel–Jordan Peace Treaty, Annex II, 1994a);
- for Jordan – 40 million cubic metres of water per year from the Jordan River (Israel–Jordan Peace Treaty, Annex II, 1994a); and,
- for Jordan – Jordan and Israel cooperate to try to find an additional 50 million cubic metres of water per year of potable water for Jordan and develop a plan within one year (from the time of the treaty signing) to do so (Israel–Jordan Peace Treaty, Annex II, 1994a). This supply for Jordan has not yet been found.

This treaty is reputed to be 'one of the most creative water treaties on record' because it has Israel 'storing' water for later transfer to Jordan (Medzini and Wolf, 2004, pp193–204).

However, Israel and Jordan are already finding it difficult to meet their water-sharing obligations. Tensions occurred in 1999 when a severe drought caused Israel to indicate that it was unable to meet its water delivery schedule to Jordan and, therefore, to raise the possibility that it would not transfer the requisite water allocation. Jordan, in turn, threatened to take 'appropriate actions' against Israel (Berland, 2000). More recently Jordan has been unable to provide Israel with its share of the Yarmouk River, possibly due to over-extraction upstream by Syria. These incidents highlight significant weaknesses in the water agreement and illustrate the need for water-sharing agreements to be able to foresee and address extreme circumstances to help mitigate the potential for conflict (Fischhendler, 2007).

Undoubtedly, Israel and Jordan will find it even more difficult to meet future treaty obligations, with the various predicted climatic changes. In particular, decreased precipitation and more evapotranspiration (and/or more extreme weather events) mean that the average storage volume in surface reservoirs could decline by as much as 25 per cent by 2100 (Israeli Ministry of Environmental Protection, 2000, p94).

The fact that there is a water-sharing agreement in place is an important factor in considering the two countries' abilities to peacefully allocate and share scarce water resources, in light of projected climate impacts, as well as population and demand growth projections, and could help to avoid potential conflict in the future. The fact that difficulties are already being faced in fulfilling commitments on both sides, however, raises some questions as to the agreement's sustainability, given these expected changes. Each country's relative ability to mitigate and adapt to climate change will also affect the treaty's sustainability.

More specifically, new demand-and-supply side water management policies are essential to help mitigate and adapt to climate change, continue to meet water-sharing obligations, reduce political tensions, and restore the Lower Jordan River. This is the case, particularly in Jordan, where rural communities and the agricultural sector are important to the support of the Hashemite Kingdom. The late King Hussein of Jordan said that 'water is the one issue that could drive the nations of this region to war' (National Environmental Trust, 2005, p19).

In other words, the overall treaty could be jeopardized due to increased political instability. In Jordan, 70 per cent of water resources are allocated to agriculture and, in Israel, 50 per cent of water resources typically are directed towards that sector. Yet, for both countries, agriculture's contribution to gross domestic product (GDP) is no more than 3 per cent. Demand-side policy changes are needed to encourage a less water-intensive form of crop production and fewer exports of such water-intensive crops from the water-poor Middle East to the relatively water-rich European nations and Gulf states. In addition, alternative investments must be made to diversify farmer incomes away from agriculture towards more economically and environmentally sustainable land uses, such as rural tourism.

The water-sharing agreement also does not contain a provision for including other riparian countries, notably Palestine, Syria or Lebanon (all of whom share the Jordan Basin), an omission that, in the future, might lead to additional controversies. Adding other riparian countries to the treaty would likely contribute to making the treaty more sustainable and to fostering broader regional cooperation. Thus, a multilateral water-sharing agreement will almost certainly be required in the future.

Israeli–Palestinian Interim Agreement

In 1995, Israel and the Palestinian Authority (PA) signed an interim peace agreement (Oslo II Accords) (Libiszewski, 1995, pp85–86).[1] Because of the already existing political tensions and the need to share increasingly scarce water resources, Annex III, Article 40 of the Interim Status Agreement was designed to address water and sewage issues by recognizing Palestinian water rights. It allocates 28.6 million cubic metres of water per year to the Palestinians for domestic consumption and recognizes that the PA will need approximately 70 million to 80 million cubic metres of water per year in the future. The Palestinians requested far more (Israeli–Palestinian Interim Agreement, 1995). Water, therefore, was ultimately left as one of five major issues to be addressed in the final status negotiations because it remains so highly contentious.

Climate change impacts will likely exacerbate difficulties between Israel and Palestine, particularly since final water agreements are not yet in place. More extreme weather events will mean that rainwater will run more quickly over the surface of the land. Consequently, much less water will be absorbed into the groundwater of the shared Mountain Aquifer, which is the main source of drinking water for Palestinians in the West Bank as well as for many Israelis. Eighty per cent of the waters of the Mountain Aquifer are currently consumed by Israel.

Overexploitation is a real concern. If the groundwater resource is over-pumped beyond the 'safe yield', this could increase the salinity of the Mountain Aquifer and affect the recharge potential, which ultimately could lead to permanent damage. And, while the PA is currently restricted in extracting water from the Mountain Aquifer without prior Israeli approval, as water resources become increasingly scarce, the necessity and likelihood of doing so will increase.

Due to domestic and agricultural needs, the PA will be seeking larger amounts of water from the Mountain Aquifer and access for the first time to Jordan River waters. The Lower Jordan River has had all of its freshwater diverted by Israel, Syria and Jordan and little more than sewage today makes its way down the river to the Palestinian West Bank. Climate change is predicted to reduce precipitation in the Jordan Valley even further. Independent and joint actions by the two parties will be needed to address climate impacts and water needs. Israel, in particular, will have to make significant reforms in this regard (Tagar, 2007). At the same time, the PA and Israel remain at odds over the issue of water rights and the basis for allocations. The fact that there is an interim agreement in place and discussions over water resources were ongoing throughout the second *Intifada*, and since, could signal an opportunity for the two parties to eventually reach a longer-term water-sharing agreement.

As noted above, a multilateral arrangement that includes Israel, Palestine, Jordan, Syria and Lebanon would likely ensure a more sustainable agreement. Third parties, such as the US and/or the European Union, should facilitate such an accord.

Syria–Jordan and Yarmouk water agreements

Jordan and Syria also have a water allocation agreement.[2] It has been renegotiated several times under circumstances detrimental to Jordan, such that the Jordanians feel that their interests have been compromised. Consequently, tensions are often high between Jordan and Syria over water allocations of the Yarmouk River and groundwater.

The fact that an agreement is in place is valuable. However, the extent of violations, and anticipated reduced water availability due to climate change, mean that such tensions can only be expected to grow and thereby call into question the agreement's sustainability.

Lack of water agreements and peace treaties: Syria, Lebanon and Israel

Currently, there are no formal agreements between Israel and Syria or between Israel and Lebanon. Both Lebanon and Syria currently have adequate water supplies (Tropp and Jagerskog, 2006). However, with projected climate impacts, including reductions in precipitation, altered rainfall distribution patterns, and increased evapotranspiration, as well as projected population growth, available water resources will decline and likely will be insufficient to meet projected demand.

For example, available water resources are expected to decline by 15 per cent for Lebanon by 2020 (Nurse, 2007). The Litani River is no longer expected to flow into the Mediterranean and reports have predicted that Lebanon will be unlikely to be able to meet local demand in the coming 10 to 15 years (Ray, 2004). With these projected changes, Lebanon will likely seek to extract more water out of the Wazzani, which is one of the tributaries of the Jordan River that is shared with Israel. This will probably lead to greater political instability between these two nations. Several years ago, Israel said that Lebanese attempts to divert water were a *casus belli* – that is, a cause of war (Deconinck, 2006). This 'incident' required third-party intervention to prevent a heightened conflict.

Syria and Israel share the resources of the Jordan River and the Sea of Galilee. Syria, like the other riparian countries of the Jordan River, is already using about 95 per cent or more of its 'annual renewable freshwater supply' (Jutro et al, 1998). Syria is expected to experience water shortages by 2020 (Jutro et al, 1998).

With 30 per cent of the waters of the Sea of Galilee originating in the Golan Heights, the return of the Golan to Syria and the water and related physical security issues at stake are intricately linked. As climate change becomes a 'threat multiplier' by making scarce water resources more so, and by tending to lead towards increased tensions over resources in the region, the lack of formal water-sharing agreements now between these countries could make the possibility of achieving such agreements in the future much more difficult and could increase the risk of future tensions or conflicts.

Socio-economic factors

Climate change's far-reaching consequences will differ greatly depending upon geography and economic conditions, with the poorest populations most likely being hit the hardest. As at least one expert has noted, 'human economic behaviour and global environmental change may pose for people with a high degree of societal and environmental vulnerability a survival dilemma' (Brauch, 2007).

Moreover, 800 million people are currently at risk of hunger, and malnutrition causes nearly 4 million deaths each year, most of them in Africa. With projected temperature increases of 2°C to 3°C, 30 million to 200 million more people will be at risk of hunger, and this figure will increase rapidly with higher temperature increases. This trend can largely be attributed to the fact that approximately 75 per cent of the poorest populations rely on agriculture for their economic livelihoods.[3] Even though some countries would benefit from a modest temperature increase, the least developed countries will not be able to adjust their crop patterns easily. Temperature increases of 3°C to 4°C would result in declines in crop yields in Africa, Western Asia and the Middle East by 15 to 35 per cent.

For example, already poor Jordanian and Palestinian farmers will have their livelihoods threatened further. In the Palestinian West Bank, up to 30 per cent of GDP is currently dependent upon subsistence agriculture. With no industry or tourism, and few job opportunities in neighbouring states, Palestinians are becoming even more dependent upon subsistence agriculture for their livelihoods. Unfortunately, less water recharge in the Mountain Aquifer will result in less water output in springs, the West Bank's main water source for agriculture. Less water for agriculture will mean that Palestinians in Gaza (assuming there is no blockade in place) will not be able to export food to Israel and Europe for hard currency – and survival.

Egyptians, too, also depend heavily upon agriculture, particularly in the predominantly rural areas. Yet, loss of productive agricultural lands in Egypt due to climate change could lead to a 20 per cent drop in wheat and maize production by 2050. Moreover, agricultural production will be under further threat, due to expected sea-level rise, for example, in the Nile Delta. To elaborate, a 0.5m climate-induced rise in sea level, for example, could displace 2 million to 4 million Egyptians by 2050 (see Figure 13.1). An anticipated 1m rise could displace 6 million to 8 million Egyptians (National Environmental Trust, 2005).

Even without taking climate change into account, Egypt is already facing water supply shortages and cannot meet its agricultural, industrial and domestic

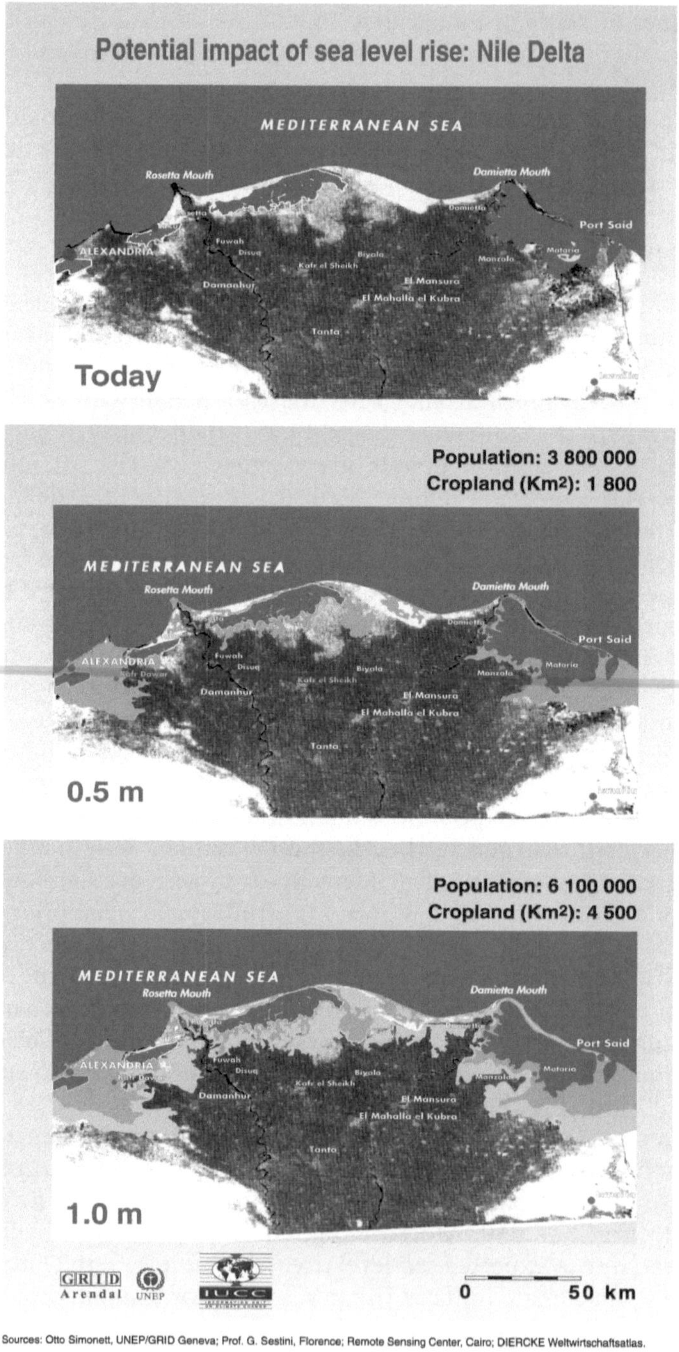

Source: adapted from National Environmental Trust *(2005)*; Simonett (2005)

Figure 13.1 *Potential impacts of 0.5m and 1m sea-level rise in the Nile Delta*

levels of demand. The creation of water and food shortages, and large numbers of displaced people and even greater unemployment due to loss of agricultural lands, industries and infrastructure from climate-induced precipitation declines and sea-level rise, will not only affect livelihoods, but also will increase competition for existing resources, which could lead to internal migration. Consequently, these factors will likely further erode public confidence in the Egyptian government, with potential for political unrest and for radical fundamentalism to grow. The Muslim Brotherhood are seen as the leading opposition and threatens the current regime's hold on power (Bensahel and Byman, 2004, pp58–59). A new regime may have a different attitude towards peace with its neighbours with drastic regional security ramifications.

Several US security experts have noted that in the developing world, even a relatively small climatic shift can trigger or exacerbate food shortages, water scarcity, destructive weather events, the spread of disease, human migration and natural resource competition. These crises are all the more dangerous because they are interwoven and self-perpetuating: water shortages can lead to food shortages, which can lead to conflict over remaining resources, which can drive human migration, which, in turn, can create new food shortages in new regions (Campbell et al, 2007, p56). In fact, the term 'climate refugees' was coined in recent years. The number of these climate refugees is estimated to reach or even exceed 200 million people by 2050 due to uninhabitable land, extreme weather events, desertification, sea-level rise and/or the salinization of agricultural lands. Along these lines, the question will arise as to whether such individuals or refugees will be allowed to enter other countries, and how these issues will be addressed (Grungras, 2007).

The projected damage in Egypt mentioned above could lead not only to internal migration but also to migration out of the country. People trying to cross political boundaries in the region could face problems and not be allowed to enter a neighbouring state. Cross-border political tensions due to the Darfur conflict and the related refugee issue already exist between Sudan and Egypt, as is the case between Israel and Egypt with thousands of Sudanese refugees now crossing from Egypt to Israel.

Conclusions and recommendations

Existing water resources in the Middle East are inadequate to meet each country's current internal agricultural, domestic and other usage, let alone to meet the needs of new transboundary water agreements or the minimum water needs of nature.

The political, economic and physical security risks that could result from the potential water shortages, due to projected climate changes, are of such a magnitude that preventive actions must be taken now to protect the security of the region.

Countries that have water-sharing agreements and/or broader 'peace' agreements in place might find it easier to cooperate towards sustainable solutions provided there is the political will to do so. Where no or only interim agreements exist, it is important to finalize such arrangements now because, with anticipated climate changes, water-sharing arrangements will only become politically more difficult to achieve. Third parties will need to facilitate bilateral and regional long-term solutions.

Other factors, such as each country's level of institutional, economic and infrastructure development, will determine the extent to which it is affected by climate and its ability to mitigate and adapt to climate change. Aid agencies therefore should tailor assistance programmes to recipient Middle East countries with climate change policy objectives in mind.

Countries will have to act domestically, with near-term and long-term planning that consists of mostly demand-side but also some supply-side water and energy management policies.

Demand management policies should be the first option to be adopted by Middle East countries. Less freshwater for agriculture, an expanded use of recycled wastewater, and alternative income support mechanisms for rural communities should become the norm. Along these lines, more domestic water conservation is critical. To help achieve this goal, measures such as government incentives for water conserving policies and technologies (e.g. waterless toilets and rainwater harvesting) are essential.

Supply-side options, such as sea water desalination, which are presently promoted throughout the region, are energy intensive, contributing to increased greenhouse gas emissions that further exacerbate climate change. Desalination technology is also not equally affordable to the different countries in the region. Although the region is blessed with sunshine, it lags behind in investments in solar power. Desalination based on solar energy could be the basis for more sustainable supply-side water management options. Cross-border cooperation for sustainable solutions that involve water conservation technology transfer and joint development of large solar fields, for example, will not only help water security, but advance political security, as well.

Without combined national, regional and international commitments to deal with the climate crisis, climate change will become the new and real threat to Middle East security with spill-over security implications for the rest of the world.

Notes

1 This agreement followed the previous Declaration of Principles of 1993, which established the Palestinian Water Administration Authority (PWA), and the Gaza–Jericho Agreement of 1994, which stipulates that all water and sewage systems and resources in the Gaza and Jericho areas shall be operated, managed and developed by the PWA. Accordingly, the PWA committed to pay Mekorot Water Co (the Israeli water company) for the costs of water supplied and for the real expenses incurred in supplying water from Israel and to the PA.

2 Syria is approximately 80 per cent dependent upon other countries for its renewable water sources, as noted by Raphaeli (1998).

3 Those living on less than US$1 per day.

References

Bensahel, N. and D. L. Byman (2004) *The Future Security Environment in the Middle East*, Report prepared for the United States Air Force, Rand Corporation, Santa Monica, CA, pp58–59, www.rand.org/pubs/monograph_reports/MR1640/MR1640.pdf

Berland, A. (2000) *The Water Component of the Peace Process between the Israelis and the Palestinians*, MSc thesis, Fletcher School of Law and Diplomacy, Medford, Chapter 4, www.transboundarywaters.orst.edu/publications/related_research/berland/berland_toc.html

Brauch, H. G. (ed) (2003) *Security and Environment in the Mediterranean: Conceptualizing Security and Environmental Conflicts*, Springer, Heidelberg, p729

Brauch, H. G. (2007) 'Impacts of global environmental change for water resources of Israel and its neighbors: New security dangers and shifting perceptions', in H. Shuval and H. Dweik (eds) *Water Resources in the Middle East: Israel–Palestinian Water Issues – From Conflict to Cooperation*, Springer, Hexagon Series on Human and Environmental Security and Peace, Berlin, vol 2, p361

Campbell, K., J. Podesta, P. Ogden, L. Fuerth and J. Woolsey (2007) *The Age of New Consequences: The Foreign Policy and National Security Implications of Climate Change*, Project of the Centre for Strategic and International Studies (CSIS) and the Center for a New American Security (CNAS), Washington, DC, p56, www.csis.org/media/csis/pubs/071105_ageofconsequences.pdf

CNA Corporation (2007) *National Security and the Threat of Climate Change*, Washington, DC, p24, www.securityandclimate.cna.org/report/National%20Security%20and%20the%20Threat%20of%20Climate%20Change.pdf

Deconinck, S. (2006) *The Wazzani Incident in the Summer of 2002 – A Phoney War?*, www.waternet.be/jordan_river/wazzani.htm

Fischhendler, I. (2007) November lecturer, Department of Geography, Head of Environmental Planning and Policy Programme, Hebrew University, Jerusalem, pers comm

Grungras, N. (2007) Israel director, Hebrew Immigrant Aid Society (HIAS), 14 October, pers comm

Israeli–Palestinian Interim Agreement (1995) *Article 40*, 28 September 1995, www.mfa.gov.il/MFA/Peace+Process/Guide+to+the+Peace+Process/THE+ISRAELI-PALESTINIAN+INTERIM+AGREEMENT+-+Annex+III.htm#app-40

Israeli Ministry of Environmental Protection (2000) *Israel National Report on Climate Change: First National Communication to the Conference of the Parties to the United Nations Framework Convention on Climate Change*, Jerusalem, p94

Israel–Jordan Peace Treaty (1994a) *Annex II: Water-Related Matters*, www.kinghussein.gov.jo/peace_annexII.html

Israel–Jordan Peace Treaty (1994b) *Article 6: Water*, www.kinghussein.gov.jo/peace_6-15.html

Jutro, P., J. Morrison and A. Wolf (1998) *Environmental Water Security: Lessons from the Southwestern US, Northwestern Mexico, and the Middle East*, US Global Change Research Program (USGCRP) Seminar, 23 November (updated 3 November 2003), www.usgcrp.gov/usgcrp/seminars/981113FO.html

Libiszewski, S. (1995) 'Water disputes in the Jordan Basin region and their role in the resolution of the Arab–Israeli conflict', *Center for Security Studies and Conflict Research*, vol 12, pp85–86, www.mideastweb.org/Mew_water95.pdf

Medzini, A. and A. Wolf (2004) 'Towards a Middle East at peace: Hidden issues in Arab–Israeli hydropolitics', *Water Resources Development*, vol 20, pp193–204, www.transboundarywaters.orst.edu/publications/Medzini_&_Wolf_04.pdf

National Environmental Trust (2005) *Global Warming in the Middle East and Central Asia*, National Environmental Trust, Washington, DC, p19

Nurse L. A. (2007) *Climate Change: Some Implications for Children's Development*, Presentation for Annual Meeting of the Consultative Group, Early Childhood Care and Development, Santiago, 16 October

Raphaeli, N. (1998) *A Briefing Paper to the Staff of the Senate Foreign Relations Committee on Potential Water Conflicts in the Middle East*, The Middle East Media Research Institute (MEMRI), Arabic News.com, 24 November, www.arabicnews.com/ansub/Daily/Day/981124/1998112415.html

Ray, P. (2004) *Multi-Objective Optimization Model for Examination of Water Reuse Potential in Lebanon*, Civil Engineering, Tufts University, Medford, 25 May, p1, www.tufts.edu/tie/pdf/ray.pdf

Simonett, O. (2005) *'UNEP/GRID – Arendal; 2 million by 2050'*, www.grida.no/publications/vg/climate/page/3088.aspx

Tagar, Z. (2007) *Nature, Agriculture and the Price of Water in Israel*, FoEME, Tel Aviv

Tropp, H. and A. Jagerskog (2006) *Water Scarcity Challenges in the Middle East and North Africa*, United Nations Development Programme (UNDP)/Stockholm International Water Institute (SIWI), Stockholm

14
At the Water's Edge: Climate Justice, Small Islands and Sustainable Development

Ambassador Stuart Beck and Michel K. Dorsey

Introduction

> *For us in the small islands, this is an issue of survival – not just for AOSIS but also for LDCs and for all developing countries. In the case of AOSIS, sea-level rise threatens our sovereignty ... there is a possibility that our nations will no longer exist in the lifetime of our children and our grandchildren.* (Delegate from the Association of Small Island States (AOSIS), at the Seventh Session of the Ad Hoc Working Group on Further Commitments for Annex I Parties under the Kyoto Protocol (AWG-KP) and the Fifth Session of the Ad Hoc Working Group on Long-Term Cooperative Action under the Convention (AWG-LCA))

For all small islands, climate change presents an enormous challenge to their collective safety, security and livelihood. For some, climate change will threaten their very existence. Nowhere is the intersection of climate change, security and human rights more apparent than in small island states and nowhere is the case for immediate action to rapidly and drastically curb carbon emissions more compelling. This chapter elaborates upon the nexus of environment and development issues that small islands must grapple with as they confront looming climate change.

There is no clear definition of what constitutes an island nation. The Alliance of Small Island States (AOSIS), an ad hoc lobbying and negotiation alliance, coordinates the most prominent group of island nations actively engaged in shaping multilateral climate policy. AOSIS commonly refers to its 43 members as small island developing states (SIDS): of the 43 total AOSIS members, only 37 are United Nations member states (Schmidt, 2005) The United Nations Department of Economic and Social Affairs (UNDESA) lists 52 small island developing states. These are broken down into three geographic regions: the Caribbean; the Pacific; and Africa, Indian Ocean, Mediterranean and South China Sea (AIMS). The criteria that define SIDS is very broad. Four AOSIS members, Belize, Guinea-Bissau, Guyana and Suriname (two of which are designated as SIDS: Belize and Guyana) are not technically islands at all.

The broader context

The total land area of most small island developing states is small relative to other countries and they can ill afford to lose land mass to rising sea level. Small size means a limited resource base. Hence, SID economies depend upon a limited number of products. The relative risk of saltwater intrusion into the groundwater supply can be pronounced for small islands.

Many small island states are located far from major trade routes, which raises the cost of imports. It is likely that small island states will become increasingly dependent upon imported food and construction materials as a result of climate change in order to compensate for less productive farming and fishing and to build the necessary infrastructure for adaptation. Remoteness also makes the provision of outside emergency relief slower and more difficult, an important consideration in a future that promises more extreme weather events. The islands within archipelago states are sometimes even remote from each other, complicating coordination and communication. The archipelagos of Micronesia cover an area the size of the continental US.

Many small island states consist of low-lying islands, making them extremely vulnerable to a rise in sea level. A single flood can cripple an entire island for weeks. With limited higher ground, people are left more exposed to flooding as well as to tropical storms and hurricanes. A few low-lying island states (e.g. the Maldives and Kiribati) are already forming contingency plans to evacuate their populations (see Box 14.1).

It is becoming increasingly clear that marine ecosystems will be especially hard hit by the effects of anthropogenic carbon emissions. Coral bleaching and ocean acidification are already causing reefs to collapse and these impacts are likely to grow worse in the future. Marine ecosystems are the basis of many small island economies, driving fishing and tourism industries. Damage to coral reefs can also jeopardize an island's food security. Coral reef damage is just a precursor to shoreline problems to come.

A large percentage of island populations and infrastructure is situated near the coast. Many islands are so small and narrow that coastal living is the only option available. The impacts of rising sea level and tropical storms are most intense near the coast.

Atoll and reef islands are made of porous calcium carbonate from ancient coral reefs and, as a result, flood from within. Many of these islands also have

Box 14.1 Maldives to buy new homeland as sea levels rise

The Maldives will begin to divert a portion of the country's billion dollar annual tourist revenue into buying a new homeland – possibly in Australia – as an insurance policy against climate change that threatens to turn the 300,000 islanders into environmental refugees, the country's first democratically elected president has said.

Source: Randeep Ramesh in the Sydney Morning Herald, 11 November 2008

poor soil for farming, which is only made worse by salt deposits left from tidal flooding.

Many SIDS are far from being wealthy countries. In fact, a number of them are among the group of least developed countries. Small island states are rarely equipped with tidal barriers and coastal reinforcements necessary to weather the higher tides of the future. Nor will they be able to pay for these enhancements to infrastructure on their own. In some cases, the cost of protecting entire islands from rising sea level and storm surges will be prohibitive.

The rise of the sea

When people think about the most devastating potential impacts of climate change, rising sea level is probably the first thing that comes to mind, and for good reason. According to the Intergovernmental Panel on Climate Change (IPCC), global sea level has been rising at about 2mm per year for the better part of the last century. The IPCC projects that sea level will rise about another 30cm to 40cm by the end of the century; however, the latest research is making these projections look overly optimistic. It is estimated that nearly one quarter of the world's population lives within 100km of the coast less than 100m above sea level, making sea-level rise a global concern.

SIDS will be among the worst hit by rising sea level for obvious reasons. A larger percentage of their populations live near the coast and their small economies are usually highly dependent upon activities centred on the ocean. Rising sea level causes many problems for SIDS, which includes, but is not limited to:

- displacement of large percentages of the population, which causes stress for not only the displaced, but also for the receiving communities;
- flooding and coastal erosion, which destroys vital infrastructure;
- saltwater inundation of agricultural land, which threatens food security and forces island farmers to switch to more salt-tolerant crops;

Box 14.2 Climate change refugees to be resettled in Papua New Guinea

The world's first climate change refugees will be relocated from their Pacific island home to Papua New Guinea by March next year.

The Solomon Islands Broadcasting Corporation says 40 families from north of Ontong Java in the Solomon Islands' Malaita Province will be relocated to Bougainville. Flooding has made parts of their islands completely uninhabitable and the islands are expected to be fully submerged by 2015. The relocation is estimated to cost the Autonomous Bougainville and Papua New Guinea government millions of dollars over the next six years. One third of the 1500 residents have refused to leave the islands.

Source: ABC News (2008)

- saltwater intrusion into vital groundwater supplies, which threatens water security and decreases the availability of freshwater for drinking, sanitation and irrigation; and
- amplification of damage from intense storms due to the weakening of natural barriers.

Many SIDS are already feeling the impact. Much of the land area of Tuvalu's nine islands is regularly inundated by rising tides every year, internally displacing many people, contaminating meagre groundwater supplies, and forcing residents to grow more salt-tolerant subsistence crops. The Maldives recently attracted media attention when its president announced that the country would be setting aside funds from its tourism industry to purchase a new homeland should the need arise (see Box 14.1).

It was reported that portions of the Solomon Islands are nearly uninhabitable due to flooding, requiring the evacuation of many families to Papua New Guinea (see Box 14.2).

According to the *Fourth Assessment Report* of the Intergovernmental Panel on Climate Change, global sea level rose at an average of 1.8mm per year from 1961 to 2003, and at an average of 3.1mm per year from 1993 to 2003. What causes sea level to rise? Since 1997, thermal expansion has contributed to 57 per cent of sea-level rise, the decrease in glacier and ice caps about 28 per cent, and the loss of polar ice sheets 15 per cent. The principal reason is, thus, the increase in average temperatures (air and ocean), which was exacerbated by anthropogenic greenhouse gas (GHGs) emissions.

Island settlements have many features in common, one of which is the high proportion of coastal settlements. In the Pacific or Caribbean islands, for instance, over 50 per cent of the population is located within 1.5km of the shore, and all major infrastructure, such as airports, or economic activities are also in floodable areas on the coast. As a result, they are highly vulnerable to a rise in sea level. Tuvalu, for example, is cut from the world up to 2 weeks a year during the high tide seasons, which completely flood the island of Funafuti and its airport strip.

Sea-level rise will affect these societies on many levels. First, coastal flooding will increase. This will trigger increased erosion of coastal land surfaces. According to scientific studies, islands with substantial human modification (e.g. modern infrastructure) are much more vulnerable to coastal erosion as the natural ecosystems have been replaced by modern structures and can no longer fulfil their ecological roles.

Since most infrastructure is on the coast, the damage from erosion and flooding is likely to be burdensome for these fragile economies. As traditional housing and building have been replaced with modern structure, the cost to repair the damage has increased. What is more, mitigation measures against sea-level rise, such as building sea walls, are extremely costly.

Another negative effect of sea-level rise is increased salinity of soils and contamination of freshwater aquifers. As a result, agricultural output, especially from traditional crops, is likely to decrease. This is a serious threat, especially since population is increasing. What is more, as saltwater enters further in the islands, freshwater lenses and freshwater aquifers will increasingly be contaminated by saltwater, leading to water scarcity. This is particularly problematic since this phenomenon will be combined with the likely decrease in precipitation. Many islands are starting to invest in desalinization plants. However, their high cost is prohibitive for many islands who cannot afford to invest in these life-saving infrastructures without outside financial help.

The predictions are not optimistic. The projected increase ranges from 0.19m to 0.58m by the end of the 21st century relative to 1980 to 1999. This increase is, of course, very problematic for islands, such as the Maldives, Tuvalu, Kiribati, and other atoll-formed nations that culminate at 3m. Tide phenomena, thermal expansion and extreme weather events will contribute to making life in these islands increasingly difficult, if not impossible.

Extreme weather

> *... a synthesis of the recent model results indicates that, for the future warmer climate, tropical cyclones will show increased peak wind speed and increased mean and peak precipitation intensities.*
> (IPCC, 2007)

While no individual storm or heat wave can be directly tied to climate change, there is little doubt that higher global temperatures will lead to more frequent and more intense extreme weather events. Fuelled by higher sea surface temperatures, the IPCC projects future hurricanes and cyclones to reach higher wind speeds and bring with them more rain. The heat waves that struck Europe in 2003 and 2006 and killed tens of thousands of people will increasingly seem less like freak natural disasters and more like typical summer weather.

Extended multi-year droughts are also expected to occur over larger portions of the globe in the coming century.

SIDS are much more exposed to extreme weather events and climate variability than most countries. Low-lying islands offer little refuge to their inhabitants from incoming storms and are especially vulnerable to powerful storm surges. Rising sea level amplifies these dangers, propagating storm damage further inland. Flooding from heavy rainfall can produce dangerous mudslides. Rebuilding damaged infrastructure is also more difficult and more expensive for small island states, taxing their limited budgets. A single storm can cripple large portions of an island's economy and leave its people at the mercy of foreign aid.

Droughts are also particularly devastating for small islands. Many islands rely on regular rainfall to recharge their limited groundwater resources. When there is too little rain (or too much at one time), these reservoirs are quickly depleted, threatening food and water security.

As average temperatures increase, scientific predictions assert that extreme events such as cyclones, droughts and floods are likely to increase in frequency and intensity. Combined with sea-level rise, these extreme weather events are expected to have mostly adverse effects on natural and human systems, especially in small islands.

The variability of extreme weather is high between different regions. The South Pacific has, in particular, seen an increased activity in cyclones, especially during El Niño years. Projections show that tropical cyclones are likely to increase in peak wind speed, which could be as high as a 5 to 10 per cent increase around 2050. For islands where traditional infrastructure has been replaced with modern and more expensive to build structures, an increase in intensity will certainly signify more damage and higher costs.

In addition to cyclones, a change in precipitation patterns is likely to occur and compromise water resources in islands. In the Caribbean region, for instance, a decrease in summer rainfall will exacerbate water scarcity. The increased average precipitation in winter months will not, however, compensate from this lack of summer rain as the islands cannot store rainwater water (which flows in the ocean, increasing sedimentation because of a decrease of forest cover). The Pacific islands will also be negatively affected as a combination of decrease in rainfall and sea-level rise will create freshwater shortages, compromising the health of the islanders and their ecosystems.

What is more, as extreme weather events increase in frequency and intensity, the economy of these islands is likely to suffer. The islands' economies, in general, revolve around natural resources. Whether tourism, agriculture or fishing, the main economic sectors will be seriously affected by the multiplication of extreme weather events. Tourism, for example, will be negatively affected by transport and communication interruptions, loss of coastal vegetation or inland forests, loss of cultural heritage by flooding, or scarce freshwater supply. These extreme events are likely to exacerbate the vulnerability of these small island nations.

Ocean acidification

> *Small islands, whether located in the tropics or higher latitudes, have characteristics which make them especially vulnerable to the effects of climate change, sea-level rise, and extreme events (very high confidence).* (IPCC, 2007)

Oceans are the main carbon sink: more than half of the CO_2 emitted since the beginning of the industrial era has been absorbed by the oceans through a complex chemical reaction at the surface of the ocean. CO_2 dissolves in water, which creates carbonic acid. From the 1790s to 1994, the concentration of CO_2 in the atmosphere has risen by 118 gigatonnes of carbon (GtC), and 53GtC from 1980 to 2005. Consequently, the CO_2 intake by oceans also increased, as did the concentration of carbonic acid. This resulted in a decrease of the average pH level: for the past 20 years, the pH level has dropped 0.02 per decade. This ocean acidification or decline in alkalinity is problematic for coral reefs: to build their shells, coral reefs create calcium carbonate ($CaCO_3$). When the water becomes saturated with carbonic acid, Ca^{2+} particles are under-saturated. The lower calcium carbonate concentration challenges shell-building processes for these species. An experiment in Biosphere 2, a self-enclosed and self-sufficient world experiment in Arizona, showed that a drop in saturation of calcium carbonate correlated with a decrease in the growth rate of these organisms.

Coral bleaching

> *Sea-level rise and increased sea water temperature will cause accelerated beach erosion, degradation of coral reefs and bleaching.* (IPCC, 2007)

Oceanic islands fall into two categories: either they are products of volcanism from a stationary magma source creating a linear chain of islands or were once joined to continental land masses and became separated because of tectonic activity. Whether products of volcanism or tectonic uplift, reefs have developed on the foundations of the islands. The coastal reef acts as a natural coastal defence structure that dampens wave energy and blocks saltwater intrusion to coastal aquifers. Since coral reefs shelter coastlines and generate sand, coral reef damage can exacerbate erosion and flooding. In addition to being a barrier, a reef provides a habitat for a diverse range of aquatic life. In terms of biodiversity, coral reefs are the tropical rainforests of aquatic ecosystems. Human populations have historically depended upon reef services such as fisheries and shoreline protection.

A coral bleaching event is triggered whenever temperature anomalies exceed a critical level. Coral reef bleaching is a phenomenon whereby corals lose their pigmentation either through loss of their symbiotic algae or a reduction in chlorophyll concentrations. As a result, corals lose their normal colourful appearance and whiten. In 1998, a year when tropical sea surface temperatures were the highest on record, coral reefs around the world suffered a severe bleaching (loss of symbiotic algae) and subsequent mortality of the reef. These events were likely induced by anthropogenic global warming, which cause steadily rising baseline of marine temperatures, combined with El Niño and La Niña events. Coral bleaching becomes more frequent and severe as the climate continues to warm because reefs are very sensitive to warm sea temperature. Even those reefs with well-enforced legal protection as marine sanctuaries, or those managed for sustainable use, are threatened by global climate change.

Food security

Many SIDS are highly dependent upon subsistence fishing and agriculture for the bulk of their food supply. Some island states have successfully provided for the nutritional needs of their people in this manner for thousands of years. Climate change is threatening these two activities and leaving small islands increasingly reliant on imported food.

Climate change threatens subsistence fishing in a number of ways. Rising sea surface temperatures can lead to coral bleaching and, in extreme cases, the death of whole coral reefs. Coral reefs are the foundation of many coastal ecosystems, without which many fisheries would collapse. Ocean acidification is also an emerging threat for coral reefs, inhibiting the ability of coral to form its calcium carbonate skeleton.

Subsistence agriculture on small islands is also threatened. Flooding due to rising sea level can leave salt deposits in the soil, damaging many traditional crops and forcing farmers to switch to more salt-tolerant varieties. Rising sea level can also lead to saltwater intrusion in the groundwater supply, damaging an important source of water for agriculture. Climate change is also likely to lead to more frequent and more intense periods of drought in some regions, with obvious implications for subsistence agriculture. Higher air temperatures are altering the length and timing of the growing season, and in some cases making the climate inhospitable to traditional crops. Increased crop stress can also be an open invitation for new disease and invasive species. Lastly, soil erosion from intense rainfall can literally wash away productive land.

Water security

Reliable access to freshwater is one of the fundamental pillars on which society is built. It is a necessary prerequisite for ensuring food security and a functioning system of public health. For small island developing states, securing adequate freshwater supplies for drinking, sanitation and agriculture is a constant challenge, and climate change is only making it more difficult.

Rainfall is the primary source of freshwater for most islands. While seemingly plentiful in some regions, rainfall is not as dependable a source of water as, for example, land-based glaciers, something that low-lying and low-latitude islands lack. Instead, SIDS must rely on surface water and groundwater supplies, which themselves are recharged by rainfall. However, these sources are threatened by climate change in a number of ways.

First, shifts in rainfall patterns are expected to increase the frequency and intensity of droughts in some regions. A single prolonged drought can have disastrous consequences on pluvial agriculture and can lead to the rapid depletion of an island's surface and groundwater resources. Second, rising sea level is leading to saltwater intrusion into the groundwater supply. This is an especially serious problem for atoll islands, which are permeable and prone to flooding from within. Saltwater intrusion has already forced some islands to switch from traditional subsistence crops to more salt-tolerant varieties. Additionally, higher air temperatures lead to higher rates of water evaporation, reducing soil moisture and decreasing the rate of groundwater recharge.

There are few good options available to SIDS. Desalinization is energy intensive and expensive. Importing freshwater to remote islands is also expensive and not practicable for supporting any meaningful agricultural activity. Building new freshwater storage facilities can help to extend supplies to a certain extent; but even they would likely prove inadequate in the face of a severe drought.

An important problem already affecting numerous coastal areas, including small islands, is saltwater infiltration in soils or salinization, especially in atoll countries such as Tuvalu, the Maldives or Kiribati. Atolls get their freshwater supplies from rainfall or groundwater (rainfall filtered in the ground). Freshwater is lighter than saltwater; a lens is formed under the atoll with freshwater on top. This reserve of freshwater is vulnerable to a decrease in rainfall (as the lens cannot replenish) and over-drilling, which can contaminate the lens water with brackish water.

As projections from the IPCC have shown, there is strong evidence that water resources and distribution of rainfall on small islands will be compromised with climate change. In Kiribati, for instance, a 10 per cent reduction in average rainfall by 2050 would lead to a 20 per cent reduction in the size of the freshwater lens. In addition, increased frequency of extreme weather events, sea-level rise and resulting land loss are likely to increase the stress on freshwater lens on atolls. For example, studies in Tarawa, Kiribati, demonstrated that a 50cm rise in sea level accompanied by a reduction in rainfall of 25 per cent would reduce the freshwater lens by 65 per cent.

These negative impacts of climate change cumulated with population increase and put the availability of freshwater resources at risk. Water quality is likely to be degraded by saltwater infiltration. This could lead to health problems related to the scarcity of freshwater and to the spread of water-borne diseases. As freshwater becomes scarce, life on islands, in general, will be more difficult to sustain. The inhabitants of the Carteret Islands in Papua New Guinea are currently suffering increased water shortage and rely on coconut water since average precipitation has decreased and their freshwater supplies have been contaminated by saltwater infiltration.

Saltwater infiltration also has severe adverse impacts upon agricultural practices. As saltwater infiltrates the aquifers and soils, many salt-intolerant traditional crops, such as taro or pulaka, die from salt contamination, which affects the traditional diet of islanders. For example, across the Pacific, taro crop is a fundamental element of islanders' diet. Because of salt contamination in soils, many island communities have been forced to relocate their plantations further inland or in higher grounds. For atolls, which culminate at less than 5m above sea level, the situation is even more precarious: some island communities (e.g. in Tuvalu) have started growing traditional crops in tin cans since they are unable to relocate to higher grounds. Other communities are switching to salt-resistant crops. However, these are expensive or difficult to obtain. Indeed, some require genetic modification to adapt to saline soils (e.g. tomato and rice), or others are not part of the traditional diet of the islanders. The implantation of such crops is, thus, modifying the diet and culture of islanders around the world.

Similarly, the import of foreign goods/foods (such as rice) to palliate the scarcity of locally grown goods from increased soil salinity has not erased all problems. Islands are often removed from major centres. As a result, these imported foods are often expensive and take time to be shipped to these islands. Therefore, the risk of food scarcity is a threat that constantly looms over island communities.

Adapting to the crisis

The role of small island states in leading the multilateral agenda for aggressive, proactive United Nations system-wide responses to climate is unique and unparalleled.

A concerted effort by a group of small island states put climate change on the United Nations agenda back in 1988, when Malta persuaded the United Nations General Assembly to recognize the potential 'severe economic and social consequences' of climate change. The Republic of Maldives followed suit by hosting the 1989 Small States Conference on Sea Level Rise, which resulted in the Malé Declaration on Global Warming and Sea Level Rise. The conference also saw the creation of an action group on climate change consisting of more than 30 small island countries, later to become the Alliance of Small Island States (AOSIS). This momentum continued through the 1992 Earth Summit in Rio de Janeiro, where AOSIS, aided by the Foundation for Environmental Law and Development (FIELD) and other non-governmental organizations (NGOs), successfully lobbied for the United Nations Framework Convention on Climate Change (UNFCCC). This success story came to a fitting conclusion when the Maldives became the first signatory to the third protocol of the UNFCCC, also known as the Kyoto Protocol.

A very similar kind of prescient acuity to the possibility for real harm from the looming threat of climate catastrophe compelled AOSIS negotiators more than two decades later, in 2009, to make the case that any mechanism for maintaining flexibility in the climate agreement should aim at stabilizing the greenhouse gas (GHG) concentration at 350ppm and global temperature rise at 1.5°C with a commitment period of 2013 to 2017, instead of the proposed 2012 to 2020 window argued by the wealthier developed countries.

AOSIS members maintained the position that concluding the next commitment period by 2017 could enable a 'post-2017' agreement to respond directly to the next full report of the Intergovernmental Panel on Climate Change, which is due in early 2014.

AOSIS representatives also made the case that the commitment of the so-called Annex I, or industrialized countries, in the Kyoto Protocol should be considered as aggregate and must be in excess of 40 per cent below their 1990 level emissions by 2020 to avoid possible irrevocable anthropogenic interference with the climate system. AOSIS members also argued that the longer-term target for Annex I countries should achieve a cut of over 95 per cent below the 1990 levels by 2050.

AOSIS called for efforts to limit temperature increase to below 1.5°C, and atmospheric concentrations to below 350ppm CO_2 equivalent (AOSIS, 2008). According to AOSIS, this will require global emissions to reduce by more than 85 per cent by 2050 (Chinese Government, 2008).

At the 2009 climate negotiations in Bonn, Germany, a leading delegate from Micronesia offered the following:

> *It is clear that 2°C is too high for small island developing states. I don't need to repeat the list of effects of 2°C ... there are some issues that no amount of adaptation funding can deal with. When a hurricane wipes out your whole country, adequate adaptation funding is very hard to come by.*
>
> *And it is clear that the latest science shows that 450ppm carries a 50 per cent risk of exceeding 2°C of global warming. It is also clear that sea-level rise will be higher than the AR4 [IPCC Fourth Assessment Report] scenario of 18cm–58cm. It is more like 50cm–140cm by 2100 [in a business-as-usual scenario]. We would see the disappearance of whole countries ... the Maldives, the Bahamas ... and if not total disappearance, what would be left will not be viable.*
>
> *We do not want to be looking at options to relocate our populations. We want to take strong mitigation action so that such a scenario is not necessary. That decision rests with the AWG-KP [Annex I Parties under the Kyoto Protocol]... Either we accept a higher stabilization level and high impacts, or Annex 1 has to do more.* (Keenan, 2009)

Collaborations: The way forward for new climate policy

Throughout the Pacific, sea-level rise has not only generated plans for the relocation of populations, but such relocations are actually in progress. Yet, small island developing states are working together and working strategically to bring forward policy changes that prevent catastrophic climate change. Small island states are building the capacity of their United Nations missions by gathering together highly trained, professional staff and working collaboratively. SIDS are creating new networks of scientific, environmental and policy experts in order to share information and coordinate activities. SIDS are devising comprehensive political strategies for advancing their environment and development agenda – in the face of looming catastrophe. The efforts of small island developing states to tackle unimaginable livelihood threatening problems, with alacrity and courage, are a beacon to the rest of the world that strategic collaborations are the key sites for ushering forth monumental changes.

References

ABC News (2008) 'Climate change refugees to be resettled in Papua New Guinea', ABC News, 5 November 2008, www.abc.net.au/news/stories/2008/11/05/2411248.htm

AOSIS (Alliance of Small Island States) (2008) *Submission on Shared Vision to the Chair's Assembly Document*, in FCCC/AWGLCA/2008/Misc.5/Add.2 (Part I), p46

Chinese Government (2008) *China's Further Submission on the Shared Vision for Long-Term Cooperative Action*, in UNFCCC document FCCC/AWGLCA/2008/Misc.5/Add.2 (Part I), p146

IPCC (Intergovernmental Panel on Climate Change) (2007) *IPCC Fourth Assessment Report*, UN, Geneva, www.ipcc.ch/ipccreports/assessments-reports.htm

Keenan, A. C. (2009) 'Covering the UN climate talks, in Bonn: AOSIS rocks it on mitigation targets', 28 March 2009, http://itsgettinghotinhere.org/2009/03/28/bonn1mitigationtargets/#more-9726

Maldives Government (2008) *Statement Delivered by the Maldives*, Chair of the LDC Group at the Opening Session of Ad-Hoc Working Group on Long-term Cooperative Action, at the Climate Poznan Conference, 1 December 2008, Poznan, Poland

Ramesh, R. (2008) 'Maldives', *The Sydney Morning Herald*, 11 November, www.smh.com.au

Schmidt, C. W. (2005) 'Keeping afloat', *Environmental Health Perspectives*, vol 113, p9

Part III
Governing Climate and Energy Instability: Avenues for Preventative Diplomacy

15
Climate Risks and Insecurities

Melinda Kimble

Introduction

Our climate is changing and the burning of fossil fuels is accelerating at a more rapid rate than ever envisioned, according to the *Fourth Assessment Report* that the Intergovernmental Panel on Climate Change (IPCC) released in early 2007. Many scientists had already suspected that this may be the case; now we have the physical evidence, which includes:

- accelerated melting of major glaciers;
- sea-level rise;
- storm frequency and intensity, especially in hurricane season; and
- droughts and consequent impacts upon flora, fauna, watersheds and ecosystems (IPCC, 2007).

These are just some of the visible signs resulting from the slight rise (0.75°C) in the average temperature since the beginning of the industrial age (circa 1750). These measurable impacts of warming further confirm that the oceans are not – as many thought in past decades – a nearly limitless carbon 'sink'. In fact, the oceans are being affected directly by the acceleration of the global water cycle, as seen with sea-level rise, increased intensity of hurricanes and typhoons, and the bleaching of coral reefs (a sign of change in the oceans' own temperature).

The IPCC report is further verification that the Earth is warming – and most of that warming is directly linked to human activity, primarily the burning of fossil fuel. The combustion of these fuels creates greenhouse gases, which trap heat close to the Earth's surface, acting like a blanket, rather than letting it rise into space. Beyond this well-accepted linkage, further evidence that warming is accelerating, is illustrated by new measurements of glacier flow in Greenland and Antarctica (ACIA, 2004). These changes imply that impacts upon sea-level rise may be larger and occur sooner than earlier models suggest.

What is harder to predict is how modern society will respond to a rapidly warming planet. In fact, for much of modern history – from the rise of civilization in the Near East – the planet has been warming and the glacial relics of the most recent Ice Age have been in retreat. During the last 1000 years, civilization has made great strides and this progress has been more dependent upon climate stability than we appreciate.

People today tend to see history as a record of steady progress and technical innovation. More recent research, however, shows this assumption to be more myth than reality. New studies indicate that climate impacts upon our ancestors were not only dramatic, but often resulted in societal collapse (Diamond, 2005). Since we are only beginning to understand the reasons for these societies' collapse, we have only brief outlines of their cautionary tales to draw on. Moreover, the most highly industrialized civilizations have a record of adaptation and adjustment to inclement weather, droughts and even sea-level rise. Given this adaptability, the most advanced societies are only now becoming fully aware of the scale and magnitude of the challenges resulting from climate change. This chapter briefly examines how climate change may compromise social, economic and political stability.

Coastal connections and resource conflicts

The global population is moving to coastal zones, according to United Nations reports. Nearly half of the world's population lives within 200km of coastline, and that number is likely to double by 2025 (Creel Population Reference Bureau, 2003). Although some of these are large inland lakes, such as the Great Lakes, much of the concentration is near the ocean. Sea-level rise will impact directly upon these areas as well as interior floodplains.

Highly developed countries, such as the UK and The Netherlands, have invested heavily in engineering solutions to protect urban areas from flooding. In the UK, the Thames Barrier is a massive engineering structure designed to reduce tidal river flow during storms. From 1990 to 2002, the system has operated, on average, four times per year. In 2003, however, it was put into operation 14 times, with two-thirds of all closures taking place since 2000 (Environment Agency, 2009). When designed, it was estimated that it would protect London from serious flooding through the 21st century – a forecast that seems overly optimistic today.

Much of The Netherlands was reclaimed from the sea and the small land area has been protected for centuries by systems of dikes and canals that were significantly strengthened after World War II. More recently, given sea-level rise forecasts, The Netherlands is exploring floating housing systems and other alternatives to adapt to the changing sea level as their engineers have determined that raising the dikes is no longer practical (Kolbert, 2006).

A sea-level rise of 1m jeopardizes nearly every coastal city – its economy, inhabitants and infrastructure. If the US response to the devastation of New Orleans by Hurricane Katrina in August 2005 is a guide, even developed countries may not be fully prepared to deal with climate impacts of this magnitude. If the forecasts are right, we will need to deal with similar events on a more frequent basis as the planet warms. To maintain urban stability and security in this environment, new strategies are needed to assess the risk of catastrophic flooding, new emergency plans are needed to move people and equipment in and out of danger zones, and new leadership is required to sustain support for a response capacity that is a quintessential public good. We may also have to abandon some areas and create new urban centres; but few policy-makers are yet considering how climate impacts may cause relocation, demand major new infrastructure investments or force a change from 'business-as-usual' patterns of behaviour.

The UK and The Netherlands have invested financial and technological resources to protect their coastal regions from the consequences of sea-level rise – and even these countries now recognize there will be a measure of flooding that they will have to accept. Few developing countries, however, have the capacity to undertake expensive engineered solutions.

In 2004, the Maldives were overwhelmed by an Indian Ocean tsunami that damaged critical infrastructure on 60 of the inhabited islands, and more than 20 of the smaller uninhabited islands disappeared. More than 1000 islands comprise the archipelago, and the average elevation is less than 3m above sea level. Engineering solutions are not an option for the Maldives or many other small low-lying island states that may well be inundated. The threats posed by climate change have encouraged this group to organize politically within the United Nations as the Alliance of Small Island States (AOSIS) and to promote early action and support for adaptation. With the most recent IPCC report, the international community can expect more urgent demands for action.

Bangladesh poses an even more complex challenge. The country is a floodplain with seven major rivers flowing into the low-lying area from the Himalayas. A sea-level rise of 1m will eliminate a significant portion of the country, displacing millions – a worst case scenario estimate suggests 55 million people could be displaced (Hulme, 2001). Today, the movement of 500,000 refugees has political, economic and social consequences; but a shift of 55 million would exceed the capacity of Bangladesh or its neighbouring states to manage and absorb. And if it occurred rapidly, in a decade – instead of a century – the population shift would severely pressure neighbouring states, leading to possible conflict.

A measure of the impacts of such human displacement exists in two recent mass migrations prompted by resource competition and long-term drought – the Rwandan genocide and the displacement of farmer communities in Darfur Province of Sudan. In the spring of 2004 more than 2 million people moved into refugee camps in the border area of Rwanda and the Democratic Republic of Congo as the genocide unfolded and more than 1 million Rwandans were internally displaced. The movement took less than 90 days as whole villages fled the rampaging Rwandan military. Yet the international response took weeks to mobilize and was barely adequate – although many more would have died without it. Rwanda illustrates the potential scope of human security challenges should massive migration result from climate-related change.

The Darfur conflict in eastern Sudan appears directly related to climate change (Faris, 2007). Rainfall patterns declined during the early 1990s, creating severe competition between nomadic herders and farmers. The Sudanese government openly sided with the nomads and their *janjaweed* militias, permitting them to force farmers off their lands. Since 2002, humanitarian experts estimate that more than 1 million people have been displaced and approximately 200,000 have died (International Crisis Group, 2007). To address the crisis, the international community has sought to establish peacekeeping operations, provide humanitarian assistance and impose sanctions on Sudan. This imperfect response has provided some protection for the refugees; but the conflict continues to simmer. International attention has been focused, justifiably, on stopping the killing; but unless the rainfall returns and the competition between herders and farmers is reduced, a sustainable solution to the crisis appears unlikely.

These conflicts – and the human consequences that ensue – offer a window on situations that could occur with greater frequency as climate change progresses. Such events further underscore the extreme vulnerability of people everywhere, but particularly in developing countries, to climate patterns that compromise traditional livelihoods. Yet, on a smaller scale, environmental migration is already under way. Hurricane Mitch in 1996 displaced entire communities in Honduras, Guatemala and El Salvador with a resultant surge in illegal and legal migration to the US (Glantz and Jamieson, 2000). Similarly, African migrants from the drought-prone Sahel continue to move into Europe. Climate change may well amplify these trends and their consequent social tensions (United Nations Foundation and Sigma XI, 2007).

Extreme weather risks

Paradoxically, as global warming raises the risk of floods, it also increases the probability of long-term droughts. Past rain and snowfall patterns are likely to change – with potentially severe impacts upon agriculture production (United Nations Foundation and Sigma XI, 2007). In fact, we may have seen the first signs of this problem during the 1970s, when the rainfall patterns that sustained sub-Saharan African agriculture in the arid Sahel changed for the worse. The ensuing decade-long drought caused cyclical crop failures, livestock deaths and famine across the Sahel – and contributed to the Ethiopian famine in the 1980s. New research suggests that this rainfall shift was most likely caused by atmospheric pollution and global warming, which shifted the rainfall north of its normal zone (NOVA, 2006). Current IPCC modelling suggests some areas in the Sahel will have more rainfall; but a larger area will have less – impacting negatively upon agriculture production – in regions that are barely achieving subsistence (United Nations Foundation and Sigma XI, 2007).

Future rainfall shifts expected in the temperate zone are likely to occur in highly productive regions like the US ‘corn belt’. US agriculture, which is concentrated in the central plains and California, accounts for 20 per cent of global production. While some of this production is irrigated, much of the cereal crops depend upon rainfall. Persistent drought conditions will reduce production and, likely, the global food balance. Globally, lower production and rising food prices could occur over an extended period as other cereal-producing regions also experience climate change. Over time, production would move to less affected zones (e.g. possibly Canada); but farmers in the US central plains may go out of business. A draft study by the US Climate Science Program forecasts a number of the consequences from climate change – lower yields, less production and higher crop loss due to pests (USCCSP, 2007).

Changes in snowfall patterns could threaten major irrigation schemes in California's Imperial Valley and other areas of the US south-west if the spring melt water flow declines. During the last century, politicians and planners saw the harnessing of rivers – the Platte, the Rio Grande and the Colorado – as opportunities to develop agriculture in relatively arid regions. These popular public investments made settlement possible and supplied hydroelectric power to the western US. In Europe, the Rhine Valley industry has been powered by similar snow-fed hydroelectric systems for decades. The Alpine snow pack is also under pressure. The prospect of reduced water availability and power suggests that the economic activity and energy demand in these regions are likely to change. If we are to sustain productivity and individual livelihoods under new climate conditions, planning must begin now.

Several climate models forecast that the agriculture situation may improve in parts of Russia and China; but some regions within both countries may suffer from production declines (United Nations Foundation and Sigma XI, 2007). The consequence of lower food production globally, combined with the disruption of climate-related natural disasters, has serious implications for national and regional security, which is dependent upon society's ability to maintain economic growth and social development. If these changes occur over long periods of time and vary from region to region, the world may have the ability to adapt to new economic development patterns. Sustained disruptions in major producing areas, however, will have global and local effects – and many of these could occur simultaneously (United Nations Foundation and Sigma XI, 2007). As the global economy's resilience to meet such challenges is predicated on maintenance of political and economic stability in the developed world, anticipating and developing a response to these challenges is a clear responsibility of policy-makers.

Biodiversity and climate

As scientists more thoroughly understand the climate cycle, ensuring resilience within ecosystems becomes vitally important. Adapting to climate change would pressure even healthy ecosystems (United Nations Foundation and Sigma XI, 2007). The United Nations Millennium Ecosystem Assessment (MEA, 2005) catalogues ecosystems under pressure across the world – and very few are deemed healthy. Watersheds and the forests that protect them are already showing some of the adverse effects of climate change. In the US Rocky Mountains, some 12,000 square kilometres of pine forests have been destroyed by a combination of fire and beetle infestation. The fires normally occur when drought has created large areas of dead timber; but the beetle infestation – directly attributed to warming – has intensified the impact (Breshears et al, 2005). The die-off in the pine forests is consistent with a US government assessment that forecasts the complete loss of 'alpine and subalpine habitats by 2100' in the contiguous US and the 'consequent extirpation of local species' (National Assessment Synthesis Team, 2000). Similar vulnerabilities to extinction are projected in unique ecosystems in South Africa and Australia (Pimm and Raven, 2000; United Nations Foundation and Sigma XI, 2007).

The large mammals in the rapidly changing Arctic also show signs of stress. Recently, the US Department of Interior confirmed scientific findings that polar bears are highly endangered (*New York Times*, 2006). Bears have proven over the last 20,000 years to be one of the most adaptable mammals. Polar bears – in developing colouration and other adaptations to a frozen world – were a remarkable example of the species' flexibility. Scientists are concerned that the speed of the current warming, however, may compromise the bears' ability to adapt.

Biodiversity in agriculture is also critical to food production, economic stability and future growth. A decline in biodiversity could be more threatening to societal survival than general species loss. The moderate and relatively predictable climate of the past 250 years has paralleled the expansion of 'industrial agriculture production', especially in the developed world and the large emerging economies. This practice has created vast 'monocultures' of grain crops, which underpin the global food system and rely on a few varieties of several grains (wheat, rice and maize) and key oilseeds (rape, olive, soy and palm). The dominance of these crops makes modern society more vulnerable to climate change. These crops comprise much of the world trade in agriculture products and the food supply of the global industrial economy. A major disease in one grain could cause widespread food insecurity and the poorest nations would be most vulnerable – much as potato blight led to the Irish famine in the 1870s.

The international community has taken some precautionary steps to protect agriculture biodiversity, most notably the International Treaty on Plant Genetic Resources for Food and Agriculture. This agreement, which is linked to the larger Convention on Biological Diversity (CBD), reflects a practical approach to protecting seeds, cultivars and other genetic material that is now held worldwide in national seed banks and in the regional research centres of the Consultative Group on International Agriculture Research (CGIAR). The US Agency for International Development (USAID) and European donors have contributed to an endowment to protect plant germplasm and established the Global Crop Diversity Trust to organize and support this effort. Recent support from the Bill and Melinda Gates Foundation and the Rockefeller Foundation has put the endowment on track to meet its initial goal of US$260 million. The trust is assembling a duplicate collection of 1 million varieties of agricultural crops and working with governments and international institutions to restore and maintain existing national and international collections. The realization of this novel idea is an example of how the international community can take creative steps to respond to the threat of climate change and support environmental sustainability. But this is only a small step.

Protecting seeds and cultivars is insufficient to address the threat that climate change poses. We will need to pursue wide-ranging, innovative research into agricultural crops – and deploy new varietals in many countries. Variety will be essential to resilience. Plant breeding – especially in vulnerable regions – is of critical importance. Out of every ten plant breeders today, only one works with 'heirloom' varieties; most focus on fewer than ten major crops and much of the research is supported by corporations that focus on supplying single varieties to large numbers of farmers.

In a future shaped by less predictable weather patterns, more pests and diseases, and scarce water supplies, crops will have to be tailored to the regional environment. New drought- and pest-resistant qualities will be increasingly important. The research investment must be made in neglected food crops that have the potential to supply better nutrition to the most marginalized people in the developing world. This transformation must also include support to farmers in growing the new crops and encouragement for mixed cropping patterns instead of the monocultures that dominate production agriculture. Without an investment in changing agricultural systems globally, the food production will be more vulnerable to climate change and populations will be more insecure. The Bill and Melinda Gates and Rockefeller Foundations are leading efforts to strengthen agronomic research in Africa. Clearly, this is an important step that must be reinforced by the public and private sector.

Sovereignty and stability

The physical phenomena of global warming are disturbing; but what is less appreciated is the potential – only recently considered – of the implications for political relations among states and how power balances might shift. The disappearance of current coasts under rising seas will have direct impacts upon the inhabitants and their sovereign states.

Sovereignty issues are an emerging issue in the climate debate. The prospect of a navigable – and exploitable – Arctic has prompted states such as Canada, Russia, Norway, Denmark and the US to assert national claims to certain areas. Canada recently declared certain border areas an 'inland sea' – and the US responded that the Arctic was an 'international waterway'. Russia, harkening back to its Czarist and Soviet traditions, planted a flag this summer near the North Pole – an action that provoked any number of claims and counter-claims. Ironically, the big prize in the Arctic is likely to be recoverable oil and gas reserves – providing a new source of greenhouse gas emissions.

At another level, fishery habitat and ocean currents may shift, complicating the ability of coastal fishermen to maintain their livelihoods. During the 1990s, the US and Canada spent more than three years renegotiating the Pacific Salmon Agreement as fishing vessels were seized and border tensions heightened. This experience may become much more common as climate change unfolds. As coastlines change, the global and bilateral agreements that govern everything from offshore fisheries to oil exploration will be tested. For example, the United Nations Convention on the Law of the Sea (UNCLOS) provides every nation with a 500 mile exclusive economic zone (EEZ). If a coastline disappears, presumably the EEZ changes – creating the potential for disputes over fishing areas and, possibly, oil and mineral deposits.

A complicating factor in the 'changing coast scenario' is the failure of the US to ratify the UNCLOS agreement. As the major superpower, the US could exploit its military and naval capability to resolve jurisdictional issues that might arise. Other states, such as the Russian Federation, might respond – leading to more tensions and possible conflicts – and this is just one possible scenario. Jurisdictional issues over well-established zones may become more frequent given the resources potentially at stake. Whether states solve these problems through dialogue and negotiation or through military engagement will shape the future.

Is international action possible?

In the two decades since the end of the Cold War, disequilibrium in global political and economic relations has become the new norm. The 'sole remaining superpower', the US, has chosen to address the imbalance through unilateral action (e.g. Iraq) and potential new – and unforeseen – alliances (e.g. the US–India nuclear deal). New bilateral trade agreements have multiplied. Russia, the European Union and China are also experimenting with a variety of new arrangements that are weakening the structures of global cooperation; but a new equilibrium is not yet in place. As a result, the risks to human security that depends upon political and economic stability are increasing.

Environmental threats have grown in urgency – the Millennium Ecosystem Assessment and the fourth IPCC report offer reams of peer-reviewed scientific studies that raise a host of cautionary flags on everything from species loss to climate change. In February 2007, Sigma XI – the Scientific Research Society – and the United Nations Foundation (UNF) released a summary report, drawing on MEA and IPCC research. This report, *Confronting Climate Change: Avoiding the Unmanageable and Managing the Unavoidable*, prepared for the 15th session of the United Nations Commission on Sustainable Development (CSD), provides a comprehensive overview of the impacts forecast and the potential for addressing these challenges. The growing scientific consensus is: start addressing climate change now as other environmental policies – conservation, habitat restoration, chemical and pollution management – are reinforced. A careful reading of the data underscores that actions in other environmental areas will eventually be undermined without a concerted global effort to stabilize and reduce fossil-fuel sourced greenhouse gas emissions. The UNF/Sigma XI report urges a variety of immediate actions to mitigate future climate change, as well as a variety of adaptation strategies to cope with the climate change that is already programmed into the system as a consequence of carbon loading of the atmosphere. Ambitious efforts to reduce global emissions are imperative, and prospects for success are uncertain.

The United Nations Framework Convention on Climate Change and its Kyoto Protocol offer a platform for reinforced global action that has been stymied, in large part, due to US opposition to the mandatory reductions in emissions required of Annex I parties to the Kyoto agreement. Yet, despite the Bush administration's opposition, individual US states, notably California and a regional alliance of north-eastern states in the US, have put in place emission cap-and-trading among eligible companies. These 'cap-and-trade' regimes parallel the design of the European Emissions Trading System. With the electoral shift in congressional power in 2006, the US started to move more rapidly at the federal level, expanding the scope for broader global action with the new administration. In a relatively short time frame, the Obama administration has strengthened auto efficiency standards and is working with the Congress and Senate to develop a federal cap-and-trade system. While positive, public support for sustained action on climate change must be strengthened and maintained as the world must sharply curb its use of coal, oil and gas between now and 2030 to have an impact upon future climate – no simple task. In the months to come we expect more aggressive action from the Obama administration.

The US – even with a strong domestic consensus for action – is not in the position it was in 1957 when President Eisenhower launched a major initiative within the United Nations to halt the spread of nuclear weapons, even though the threat is comparable. With India, China and Brazil growing at unparalleled rates, economic dynamism has expanded far beyond the industrial West. China and India combined already consume more power than the European Union and their domestic demand will only continue to grow – and both countries are currently heavily committed to using domestic coal. Developing countries contend, with some justification, that developed countries caused the problem and the wealthier states must take action. Had the international community accepted this proposition and initiated action in 1992 when voluntary commitments were adopted to return greenhouse gas emissions to 1990 levels by 2000, we would be in a better position to pursue step-by-step incremental action.

Today, with global emissions more than 25 per cent above 1990, action by every major emitter is essential. US participation – in concert with strong actions already under way in the European Union – would add needed momentum to international effort; but such a move must be met by action in China, India, Brazil and other large emerging economies that would reinforce the new alignment. To encourage developing countries to act, the international community will need to find ways of deploying the most efficient technology as quickly as possible in the developing world. Given the widespread use of coal, new global standards are needed for coal-fired plants and the cleanest power systems used; but new incentives are required globally to ensure that coal for power generation is no longer a 'first, best' option in many countries. To achieve such an objective requires more innovation and deployment of renewable energy – and urgent research on carbon capture and storage for coal plants that may ultimately be built.

All of these actions demand a new commitment to international cooperation and a reinforcement of the institutions – the United Nations, the World Bank Group, the Global Environment Facility (GEF), and the regional development banks – that support and finance many initiatives. Most importantly, the private sector with its resources, innovation and ingenuity must be engaged globally and markets must be structured to send signals that drive change. The greatest risk is that we will fail to act quickly enough to 'avoid the unmanageable impacts' of climate change.

References

ACIA (2004) *Impacts of a Warming Arctic: Arctic Climate Impact Assessment, 2004*, Cambridge University Press, www.acia.uaf.edu

Breshears, D. D., N. S. Cobb, P. M. Rich, K. P. Price, C. D. Allen, R. G. Balice, W. H. Romme, J. H. Kastens, M. L. Floyd, J. Belnap, J. J. Anderson, O. B. Myers and C. W. Meyer (2005) 'Regional vegetation die off in response to global change type drought', *Proceedings of the National Academy of Sciences*, www.pnas.org/content/102/42/15144.abstract?ck=nck

Creel Population Reference Bureau (2003) *Ripple Effects: Population and Coastal Regions*, Population Resource Board, www.prb.org

Diamond, J. M. (2005) *Collapse: How Societies Choose to Fail or Survive*, Penguin Books, New York, NY

Environment Agency (2009) 'Thames Barrier closures', http://www.environment-agency.gov.uk/research/library/publications/41065.aspx

Faris, S. (2007) 'The real roots of Darfur', *Atlantic Monthly*, April, pp67–69

Glantz, M. and Jamieson, D. (2000) 'Societal Response to Huricane Mitch and Intra versus Intergenerational Equity Issues: Whose Norms Should Apply?' Risk Analysis, vol 20, issue 6, pp869–882

Glantz, M. H. and Jamieson, D. (2001) *Climate Ethics and Equity*, Environmental and Societal Impacts Group for the National Center for Atmospheric Research, National Science Foundation, Boulder, CO

Hulme, M. (2001) *Can Britain adapt to Climate Change?*, Tyndall Centre for Climate Change Research Conference, UK

International Crisis Group (2007) *A Strategy for Comprehensive Peace in Sudan*, Africa Report N130, www.crisisgroup.org

IPCC (Intergovernmental Panel on Climate Change) (2007) *Climate Change 2007: Impacts, Adaptation and Vulnerability: Contribution of Working Group II to the Fourth Assessment Report of the Intergovernmental Panel on Climate Change*, Cambridge University Press, Cambridge, UK

Kolbert, E. (2006) *Field Notes from a Catastrophe*, Bloomsbury, New York, NY

MEA (Millennium Ecosystem Assessment) (2005) *Ecosystems and Human Well Being: Our Human Planet. Summary for Decision Makers*, Island Press, Washington, DC

National Assessment Synthesis Team (2000) *Climate Change Impacts on the United States: The Potential Consequences of Climate Variability and Change*, Cambridge University Press, UK, and US Global Change Research Program, Washington, DC

New York Times (2006) 'Agency proposes to list polar bears as threatened', *New York Times*, 28 December

NOVA (2006) *Dimming the Sun*, A DOX Production for NOVA/WGBH and BBC, WGBH Educational Foundation, April 2006

Pimm, S. L. and P. Raven (2000) 'Extinction by numbers', *Nature*, vol 403, pp843–884

United Nations Foundation and Sigma XI (2007) *Confronting Climate Change: Avoiding the Unmanageable and Managing the Unavoidable*, Scientific Expert Group Report on Climate Change and Sustainable Development, Prepared for the 15th UN Commission on Sustainable Development, UNF, Washington, DC

USCCSP (US Climate Change Science Program) (2007) *The Effects of Climate Change on Agriculture, Land Resources, Water Resources, and Biodiversity*, USCCSP, Washington, DC

16

A Multilateral System for Climate and Energy Security: What Roles for Existing Institutions?

Felix Dodds and Richard Sherman

> *Our lives on this planet are too short; the work to be done is too great. But we can perhaps remember that those who live with us are our brothers; that they share with us the same short moment of life.* (Senator Robert Kennedy, Jnr, 1968)

The present state of play

The United Nations General Assembly mandate establishing the Brundtland Commission in 1982 encouraged the commissioners to look for 'long-term environmental strategies for achieving sustainable development to the year 2000 and beyond' (Brundtland, 1982).

Unfortunately, although we have made great strides in the last 20 years, it is not nearly enough. There have been many broken promises by developed countries, which have failed to make available the resources to enable developing countries to take a more sustainable development model. Also, developed countries have not adequately addressed their own consumption patterns, or implemented policies based on the principle of 'common but differentiated responsibilities' that underpin both the Rio and Johannesburg Earth Summits. In doing so, they have put the world under threat.

In 2007, an environmental issue reached the United Nations Security Council for the first time. That, of course, was energy and climate security. Brought there by the UK government, it is the topic of this book. But it wasn't the only environmental issue to be discussed that year in the Security Council.

Two months later, Belgium also brought an environmental issue to the Security Council – that of reviewing 'peacekeeping mandates to consider helping states prevent illegal exploitation of natural resources from fuelling conflicts'. What is happening is an increasing convergence of the environmental, human development and security agendas. A key question, then, is: is the Security Council the right place to discuss these issues?

The Security Council has what is known as 'the permanent five members' (China, France, the Russian Federation, the UK and US). These countries play a significant role in causing the world's environmental and development problems, and have a veto on possible actions. They account for around 60 per cent of global carbon dioxide emissions. We argue that the Security Council is not the place to discuss these issues ... yet. It may become an action of last resort if negotiations fail through other multilateral processes.

The past few years have seen climate change become the single most important issue on governments' environmental agendas, often to the detriment of their work on other issues. Governments' environmental departments have seen their budgets for non-climate-related work shrink.

We raise a serious note of caution about this too. Putting everything under the banner of 'climate change' is not a solution to the problems we are facing. It is critical that we remember that climate change will definitely make things worse; but things are already going in the wrong direction. We need to address other priority environmental and development issues, along with addressing climate change. It is unlikely that more than a handful of the Millennium Development Goals (MDGs) will be fulfilled, and even those that were on track are now under threat with the present economic crisis. MDG 7 on environmental sustainability didn't adequately address the challenges we face in the area of environmental threats.

The excellent United Nations Environment Programme (UNEP) publication *Global Environmental Outlook 4* (GEO4) (UNEP, 2007a) did point out the key trends that we really do need to address:

- The world's population has grown by 34 per cent to 6.7 billion in 20 years.
- Human consumption has far outstripped available resources. Each person on Earth now requires one third more land to supply his or her needs than the planet can supply.
- Irrigation for agriculture already takes about 70 per cent of available water; yet meeting the Millennium Development Goal on hunger will mean doubling food production by 2050. By 2025, water use will have risen by 50 per cent in developing countries and 18 per cent elsewhere.
- Sixty-four per cent of ecosystems are degraded and are being used unsustainably.
- Over 73,000 square kilometres of forest are lost across the world each year – 3.5 times the size of Wales.
- An estimated 75,000 people each year are killed by natural disasters.

- Three million people die needlessly each year from water-borne diseases in developing nations – mostly children under five.
- Sixty per cent of the world's major rivers have been dammed or diverted.
- Populations of freshwater fish have declined by 50 per cent in 20 years.
- More than half of all cities exceed World Health Organization (WHO) pollution guidelines.
- Biodiversity is seriously threatened by the impact of human activities: 30 per cent of amphibians, 23 per cent of mammals and 12 per cent of birds are under threat of extinction, while one in ten of the world's largest rivers runs dry every year before it reaches the sea (UNEP, 2007a).

The report further states that land degradation affecting about 5 million square kilometres of Africa – one sixth of the continent – is the biggest threat to the region realizing its full potential.

As Achim Steiner, the executive director of UNEP, said:

> *But all too often [the response] has been slow and at a pace and scale that fails to respond to or recognize the magnitude of the challenges facing the people and the environment of the planet... The systematic destruction of the Earth's natural and nature-based resources has reached a point where the economic viability of economies is being challenged – and where the bill we hand to our children may prove impossible to pay.* (Steiner, 2007)

The GEO4 report underlined the findings in the Intergovernmental Panel on Climate Change's (IPCC's) Fourth Assessment Report when it said: 'irreversible damage to the world's climate will be likely unless greenhouse gas emissions drop to below 50 per cent of their 1990 levels before 2050. To do this, it [the IPCC)] said that developed countries must cut emissions by 60 to 80 per cent by 2050 and developing countries must also make significant reductions'. (UNEP, 2007)

In response to the report David Nussbaum, the chief executive officer (CEO) of the World Wide Fund for Nature-UK (WWF-UK) pointed out: 'If everyone consumed resources in the way that we do in the UK or Europe, we would need three planets to support us' (Nussbaum, 2007). And this by 2030!

The world of climate change: United Nations agencies and programmes dealing with climate change

Outlining the larger problem we face might perhaps help us to focus on the problems in the area of energy and climate. What has been very clear since the Rio Earth Summit is that governments are not very good at joined-up thinking or acting, for that matter. Perhaps this is because climate change impacts upon so many different ministries.

The inability of governments to work in a joined-up manner is equally seen in the United Nations' activities. Perhaps not everyone realizes that all UN agencies and programmes have their own governing boards made up of representatives from those same government departments, and often reflecting the lack of joined-up thinking there.

Visiting the online gateway to the UN system's work on climate change can be mind blowing, particularly if you are not aware of how extensively the UN system is involved in trying to address climate change. Its web page (www.un.org/climatechange) lists the following UN bodies involved with activities in the area of climate change: the United Nations Framework Convention on Climate Change (UNFCCC); the IPCC; UNEP; the World Meteorological Organization (WMO); the United Nations Commission on Sustainable Development (CSD); the United Nations Food and Agriculture Organization (FAO); the Global Environment Facility (GEF); the United Nations Development Programme (UNDP); the World Bank; the Convention on Biological Diversity (CBD); the United Nations Convention to Combat Desertification (CCD); the International Maritime Organization; UN regional commissions; the International Monetary Fund (IMF); the World Tourism Organization; the United Nations Educational, Scientific and Cultural Organization (UNESCO); the World Health Organization (WHO); the World Food Programme (WFP); the United Nations Human Settlements Programme (UN-Habitat); the United Nations Conference on Trade and Development (UNCTAD); the Global Climate Observing System; the United Nations Department of Economic and Social Affairs (UNDESA); the United Nations Industrial Development Organizations; the International Fund for Agricultural Development; the International Strategy for Disaster Reduction; the International Telecommunication Union; and the United Nations Institute for Training and Research.

Keeping abreast of developments within each body has been made easier by the excellent companion site produced by the International Institute for Sustainable Development: http://climate-l.org/.

As we move forward to Copenhagen and beyond, one of the key tests for the United Nations and the UN Secretary General will be trying to create joined-up thinking and activity within the UN system. This will be even more important at the country level, and the development of the UN One Country Programmes does offer a more coordinated way forward for the United Nations as a whole.

In the following sections of this chapter, we focus on the work of just a small number of the UN bodies involved in climate change: the UNFCCC, UNEP, UN-Habitat, UNDP and the FAO. These bodies, in particular, play a significant role, and understanding what they are doing can help to map out a more integrated approach. We will return to some suggestions for governments to consider in the final section.

United Nations Framework Convention on Climate Change (UNFCCC)

The first Intergovernmental Report on Climate Change in 1990 did not have the certainty of the more recent 2007 report. It produced compelling, but by no means conclusive, evidence that greenhouse gases produced by human activity are having an impact upon global climate. The report had been preceded by a meeting of governments in Toronto in 1988, where they committed to stabilizing carbon dioxide by 2000 on 1990 levels and further reducing CO_2 by 20 per cent, by 2005, on 1990 levels. There had been some hope that the 1992 convention would therefore be more than a framework convention and would, in fact, have targets. The omission of targets may be looked at by future generations as a severe lack of political leadership – one of the most severe shown by that generation of politicians.

The successful agreement of the Kyoto Protocol in 1997 can probably be traced back to the United Nations Special Session of the General Assembly (UNGASS) to Review and Appraise the Implementation of Agenda 21. The meeting played out as an informal discussion place for many heads of state prior to Kyoto that enabled the protocol to be negotiated.

There is clearly confusion on what issues the UNFCCC should take up and what should be left to others. It's not clear to many how the jigsaw of climate change institutions should come together.

The role of the UNFCCC is to:

- coordinate information-sharing about mitigation and adaptation for parties of the UNFCCC;
- support the Clean Development Mechanism (CDM) Executive Board and Joint Implementation Supervisory Committee (JISC);
- coordinate the Technology-Transfer Clearing House;
- run the Climate Change Information Network;
- support the work of the Subsidiary Body for Implementation (SBI), which develops recommendations to assist the Conference of the Parties (COP) in reviewing and assessing implementation of the convention and the Kyoto Protocol;
- support the work of the Subsidiary Body on Scientific, Technical and Technological Advice (SBSTTA), which advises on technical and scientific aspects of the UNFCCC's work.

So the UNFCCC ensures the space for governments to set targets and agree activities, and acts as a knowledge management hub.

United Nations Environment Programme (UNEP)

> *UNEP's mission is to provide leadership and encourage partnership in caring for the environment by inspiring, informing and enabling nations and peoples to improve their quality of life*

> *without compromising that of future generations.* (UNEP Mission Statement, www.unep.org/Documents.Multilingual/Default.asp?DocumentID=43)

UNEP's mandate was expanded and rearticulated by the 1997 Nairobi Declaration 'to be the leading global environmental authority that sets the global environmental agenda, that promotes the coherent implementation of the environmental dimensions of sustainable development within the United Nations system and that serves as an authoritative advocate for the global environment' (UNEP, 2007b).

It's not surprising, therefore, that UNEP has played an important role in the recent history of climate change. With the WMO, they set up the IPCC at the second World Climate Conference in 1988. Mustapha Tolba, the executive director at the time, understood the need for science to underpin political activity. Without his leadership on this, present activities on climate change would not be possible.

UNEP does not generally have national programmes in countries, although it does undertake capacity-building in areas such as the implementation of international environmental law. There has been a growing call from developing countries for more national activities under the Bali Strategic Plan. In-country environmental activity is usually undertaken by other parts of the UN, such as the UNDP.

On climate change, UNEP is already active in developing activities with the FAO and UNDP on the reducing emissions from deforestation in developing countries (REDD) initiative under the UNFCCC.

In its new *Medium-Term Strategy 2010–2013*, UNEP has identified climate change as one of the six cross-cutting thematic priorities. It sees its objective to 'strengthen the ability of countries to integrate climate change responses into national development processes' (UNEP 2008).

UNEP is active in the area of adaption. Projects include facilitating better local climate data collection and helping to determine possible impacts of climate change; contributing to improved scientific methods and assessment tools; and supporting the science and policy community in undertaking adaptation planning, including that linked to disaster prevention efforts. In the area of mitigation, it aims to help countries make sound informed energy supply choices with a focus on **renewable energy** options. UNEP also supports the development of economic activity to address climate change, such as the role of the **carbon market** in both developing and least developed countries.

It also plays an important role in helping to build public awareness of environmental issues, including climate change.

United Nations Human Settlements Programme (UN-Habitat)

Some say that the battle for and against climate change will be won or lost in the cities of the world. Doje Cering, the former Chinese minister of civil affairs, stated that in the coming two decades 'about 12 million people from China's rural areas will move to urban areas annually. Therefore, 20 new cities need to be set up (Cering, 2000).' China estimates that within a decade in China, urbanites will outnumber peasants. It is worth noting that in 1990, only 14 per cent of the world's population lived in cities. Now it is over 50 per cent, and as populations grow, the proportion based in cities will increase. This will lead to an increased impact upon the climate, particularly if new cities are not built as very low-carbon cities.

According to the C40 Initiative (a grouping of key cities around the world to tackle climate change), cities consume around 80 per cent of the world's energy and cause around 75 per cent of CO_2 emissions.

As the main United Nations body dealing with human settlements, UN-Habitat has developed the Cities in Climate Change Initiative (CCCI) as part of its global Sustainable Urban Development Network (SUD-Net). This initiative by UN-Habitat promotes dialogue between different levels of government, linking local action plans with national action plans within the framework of international agreements on climate change. This learning network is critical to addressing the challenges that cities face, both in the areas of mitigation and adaptation.

As the CCCI process starts to develop a common understanding of the issues and challenges on cities and climate change, it will need the support of governments. Governments can support their cities by enacting policy changes and activities that bridge the research and practice divide, and do so quickly.

United Nations Development Programme (UNDP)

> *Climate change is the defining human development issue of our generation. As the United Nations' global development network, UNDP's goal is to align human development and climate change management efforts by promoting mitigation and adaptation activities that do not slow down, but rather accelerate socio-economic progress.* (Vandeweerd, 2008)

The UNDP as the main development agency of the United Nations has offices in 135 countries and acts as the coordinator for the growing UN One Country programmes. Its budget dwarfs other UN bodies at around US$12 billion for the 2008 to 2009 biennium. Often involved with infrastructure projects, the UNDP is well placed to ensure that the types of development funding available through its offices are ones that only promote low-carbon economies.

Through its role in the Global Environment Facility, the UNDP has, over the last 15 years, helped to mobilize over US$2 billion to fund greater energy access and security, energy efficiency and clean development projects in over 100 countries. It has supported the ability of countries to access carbon markets, particularly the Clean Development Mechanism and joint implementation (JI). To date, the UNDP has implemented CDM and JI capacity development activities in over 20 countries. It also established the Millennium Development

Goal of Carbon Facility to assist developing countries in leveraging carbon finance for clean energy development and sustainable land-use practices.

The UNDP has developed substantial information resources on climate change and development, including the Adaptation Policy Framework, country climate risk profiles and portfolio-screening methodologies. These will be important in helping to build developing countries' ability to embed and integrate resilience-building and adaptive capacity within domestic policies and investment decision-making processes and practices. Increasingly, as climate change impacts affect the world, the UNDP network of disaster risk reduction specialists will play a significant role.

United Nations Food and Agriculture Organization (FAO)

> *In the area of climate change, the Food and Agriculture Organization (FAO) contributes to the debate by assessing the available scientific evidence, participating in observing and monitoring systems, collecting unique global datasets, promoting adaptation and mitigation practices, and by providing a neutral forum for negotiations and technical discussions on climate change and agriculture.* (FAO, 2008)

Around 30 per cent of the current contribution to greenhouse gas emissions comes from the forestry and agriculture sector. Forest degradation and deforestation account for around 17.4 per cent of GHGs, and agriculture about 13.6 per cent.

As population increases, emissions from these two sources are also likely to increase. The FAO can play an important role in providing better information on how to reduce the contributions to climate change from both agriculture and forestry. In 2008 we saw how the move to biofuels by a number of countries increased food prices. Biofuels is a complicated and interwoven problem that needs careful discussion and a commitment to sustainable development criteria before moving forward. The FAO can play a leading role here.

The REDD initiative, managed by the UNDP and UNEP, is an interesting initiative that utilizes the UN country teams and the unique roles and capacities of the partners. It hopes to build a knowledge base of what is happening on the ground. This requires an agreement on the development and testing of standards, methods and guidelines, accounting, reporting and verification.

The FAO has also been mainstreaming integrated climate change adaptation and mitigation strategies within agriculture, fisheries and national food security plans and programmes.

United Nations central

With the UNFCCC and UN agencies and programmes already active on climate change, what role does that leave the UN in New York? Well, quite a lot actually. In this section, we investigate what the responsibilities of the key bodies are.

These are the United Nations General Assembly (UNGA); the Economic and Social Council (ECOSOC) and one of its functional commissions, the Commission on Sustainable Development (CSD); the United Nations Security Council; the Secretary General's Office; and the interagency group the United Nations Chief Executives Board (CEB).

United Nations General Assembly (UNGA)

> *The General Assembly may call the attention of the Security Council to situations which are likely to endanger international peace and security.* (Article 11 of the UN Charter)

In 2002, the UN General Assembly adopted sustainable development as a key element of the overarching framework for the UN, particularly for achieving internationally agreed development goals, including those contained in the Millennium Declaration.

The World Summit 2005 *Outcome Document* also committed member states to promote the integration of the three components of sustainable development – economic development, social development and environmental protection – as interdependent and mutually reinforcing pillars.

The General Assembly has not yet been able to show how it has mainstreamed sustainable development within its actions. This has not been helped by the weakening of the Commission on Sustainable Development over the past few years. There has been a lack of vision on how UNGA should approach sustainable development, and it is continuing as it has done in the past. The decision to hold a debate on climate change in 2007 was an important move by the assembly in mapping out a role for the future.

United Nations Security Council

> *In order to promote the establishment and maintenance of international peace and security with the least diversion for armaments of the world's human and economic resources and agree that in carrying out its duties under this responsibility the Security Council acts on their behalf.* (Article 24 of the UN Charter)

The two debates on environment in 2007 were the first but probably will not be the last. The G77 and some NGOs such as Stakeholder Forum argued that the issue of climate and energy security should first be debated within the Economic and Social Council and its subsidiary bodies, and then the UN General Assembly, before it is taken to the Security Council.

The present Security Council has five permanent members, each of which has a veto. George Bush senior said at the 1992 Earth Summit: 'The American *way of life is not negotiable.*' This position makes discussion on energy and climate security problematic, to say the least. If other bodies within the UN cannot deal effectively with the gap between such rhetoric and the action required to reverse climate change, then the issue could end up being addressed in the Security Council as energy plays such a critical role to the US way of life.

United Nations Economic and Social Council (ECOSOC)

> *It may coordinate the activities of the specialized agencies through consultation with and recommendations to such agencies and through recommendations to the General Assembly and to the members of the United Nations.* (Article 63 of the UN Charter)

The UN Economic and Social Council has a central role in receiving the reports from the UN specialized agencies and from the functioning commissions. It is therefore hardly surprising that they have been given the role as the central mechanism for system-wide coordination in the UN system. ECOSOC has tended to deal with climate change in the context of the MDGs and reports from UNEP.

Commission on Sustainable Development (CSD)

> *3(h) To consider, where appropriate, information regarding the progress made in the implementation of environmental conventions, which could be made available by the relevant conferences of parties.* (UN, 1992)

The 2002 World Summit on Sustainable Development reaffirmed the Commission on Sustainable Development (CSD) as the highest-level commission dealing with sustainable development issues within the UN system.

Chapter 9 of Agenda 21 deals with climate change under the topic of 'Protection of the atmosphere'. Energy was added to the topics of the CSD in 1997 at Rio +5. Recognizing that some 2 billion people have no access to modern energy services is clearly a sustainable development issue.

It was clear to the government negotiators for both Agenda 21 and the Johannesburg Plan of Implementation that the CSD should continue to have a role in looking at both climate change and energy. Perhaps with energy it is understandable as there is no obvious other home for that discussion such as water. Recognizing this within the UN system, the secretary general set up the interagency coordination mechanisms to deal with how the UN should approach these subjects (e.g. UN Water and UN Energy). This stills leaves a massive gap in how and where governments should address energy.

When the CSD last discussed energy and climate change in 2007, the CSD for the first time was unable to come to a policy agreement. The controversy was primarily over what role the commission should play on climate change. The CSD did not, as requested by G77 and a number of countries in the Security

Council, come forward with suggestions relating to energy and climate security issues.

Farukh Amil, the deputy permanent representative for Pakistan and chair of the G77 developing countries, made some powerful arguments against the inclusion of climate change in the Security Council. While recognizing the importance of the issues for the achievement of sustainable development, developing countries felt the responsibilities to address the climate and security nexus were the responsibility of the General Assembly and the Economic and Social Council.

Clearly the CSD, if it is to continue, needs to map out much more plainly what role it could have in energy and climate change policy and action.

United Nations Chief Executives Board (CEB)

In 2003, the UN Chief Executives Board adopted a set of guidelines to support sustainable development within the UN system at the agency level, which has subsequently led to the establishment of UN Water, UN Oceans and UN Energy.[1]

One of the key subsidiary bodies of the CEB is the High-Level Committee on Programmes (HLCP). In addition to providing the forum for interagency dialogue in the development and launching of new programme initiatives, the HLCP advises the CEB on issues of 'strategic planning, policy and programme development and implementation, and on other areas that require priority attention in response to the challenges facing the UN system and the global community' (UN, 2000).

The committee also fosters and supports the integrated and coordinated implementation and follow-up of major UN conferences and summits. To enable the CEB to be effective on system-wide activities on climate change, it has established a working group on the issue. The working group reports to the HLCP, prepares input to the UNFCCC meetings and reviews the outcomes of the system as a whole.

Office of the UN Secretary General

The UN Secretary General outlined his priorities when he took office in January 2007. One of his priorities was the issue of climate change. He has consistently used his office to promote dialogue and action on climate change with some of the less than enthusiastic government leaders. On 1 May 2007, he announced three UN special envoys on climate change: Gro Harlem Brundtland, the former prime minister of Norway and former chair of the World Commission on Environment and Development (WCED); Han Seung-soo, former minister of foreign affairs of the Republic of Korea and former president of the United Nations General Assembly; and Ricardo Lagos, former president of Chile. He later appointed two more: Festus Mogae, former president of Botswana, and Srgjan Kerim, the former president of the UN General Assembly and former Yugoslav Republic of Macedonia. The role for the envoys will be engaging in consultations with governments and other organizations in order to assist the secretary general to progress the international negotiations towards a post-Kyoto climate change treaty.

The September 2007 high-level event on climate change, initiated by the UN Secretary General, set in motion the impetus for leaders to look ahead to the discussions on the UN Framework Convention meeting in December 2007. It sent the message that this is no longer business as usual.

The secretary general has also set up a unit within his office to work on climate change: another indication that he intends to ensure that the UN system as a whole can play a role in climate change.

International financial institutions (IFIs)

There is clearly substantial activity on climate change; but if it is to be successful, then it must be backed up by funds. In 1992, Secretary General of the Rio Earth Summit Maurice Strong was asked how much it would cost to implement Agenda 21. The estimate was that it would cost US$625 billion a year, of which US$125 billion would be transferred from developed to developing countries (at that time, the amount being transferred in official development aid was US$55 billion – 33 per cent of gross national income (GNI)). In 2007, it is estimated at US$103.5 billion – 0.28 per cent of GNI, less than in 1992. Perhaps he should have been asked what the cost of inaction was. We are sure it would have been more by an order of magnitude.

The fragmented nature of the multilateral environment system can be laid at the doors of governments. It has been estimated that there are over 700 Multilateral Environmental Agreements (MEAs), and although there have been moves in recent years to try and cluster conventions, this has had limited success. For small developing countries this has been a disaster, as the costs of reporting and attending a fragmented system are huge.

A more coherent and equitable approach to funding climate change mitigation and adaptation needs to be developed. What we really do not need is a plethora of funds being set up, each with their own secretariats, application processes and reporting processes.

The following sections examine a number of the funding mechanisms available under multilateral institutions. We are not looking at the Italian, Dutch or Danish funds even if they are administered by the World Bank. We look at the World Bank, the Clean Development Mechanism, the Global Environment Facility and within it the Least Developed Countries Fund, the Adaptation Fund and the Special Climate Change Fund and joint implementation.

World Bank

Before commenting on the fund within the World Bank, it is worth pointing out that as the major funder with the regional development banks, it provides financial and technical assistance to developing countries for development programmes (e.g. bridges, roads, schools, etc.) with the stated goal of reducing poverty. It does this under the Poverty Reduction Strategies and the Comprehensive Development Frameworks. Governments seeking funding grants, loans and guarantees do not have any forecasting for the impacts of the project in contributions to greenhouse gases. In addition, its funds are not yet dependent upon showing that development programmes are linked to climate reduction strategies.

Currently, projects funded by one part of the World Bank or government agency could undermine the activities of other parts of the Bank or development agency in their attempt to get the production of greenhouse gases under control. Clearly, some joined-up thinking is needed here. The World Bank Framework on Climate Change does nothing to address this.

The battle between the World Bank and the GEF for control of most of the climate funds in the future needs real debate, as does where future funding for climate change will fit in the much broader international environmental governance landscape.

Finally, the funds so far available from bilateral means (e.g. Japan Cool Earth Partnership, the Environmental Transformation Fund – International Window (ETF-IW), the North American Aerospace Defense Command (NORAD) Rainforest Fund, the European Commission's Global Climate Change Alliance (GCCA), the Spanish MDG Fund and the German International Climate Initiative) are of a magnitude of around five times that available from the multilateral agencies, although some of the money at present channelled through the bilateral funds may be committed to multilateral funds in the future.

World Bank Climate Investment Fund (CIF) The World Bank and other members of the multilateral development banks (MDBs) have developed the Climate Investment Fund (CIF) as an interim measure. This is to help scale up assistance to developing countries and thus strengthen the knowledge base in the development community on climate change adaptation and mitigation. In 2008 the donors approved US$6.1 billion, although it is unclear where this is going to come from as of yet. The CIF is broken down into two separate funds: the Clean Technology Fund (CTF) and the Strategic Climate Fund (SCF).

The CTF has a focus on financing transformation actions by offering incentives for low-carbon development and mitigation of GHGs, scaling up and sharing clean technologies, and looking for environmental and social co-benefits contributing to sustainable development and the delivery of the MDGs.

The Strategic Climate Fund is also focused on scaling up but looks at new approaches. It focuses on areas such as sustainable forest management, greening energy access and climate resilience schemes.

This represents a new source of funding that will enable the MDBs to provide additional grants and concessional financing. One interesting development is the governance model for the funds. Under the Climate Investment Fund, the decision-making committees for both the CTF and the SCF have equal members from both the donors and recipients. The role of stakeholders in advising and participation is still to be agreed.

Carbon Partnership Facility (CPF) This is more focused on fostering strategic interventions in sectors, as opposed to individual projects. It aims to link through partnership buyers and sellers, and targets long-term emissions, including in uncertain markets. It hopes to have a carbon fund of around 5 billion Euros over five years. In addition, a carbon asset development fund will be established to help provide sellers and host countries with grants or resources to enable programme development and methodological emission reduction work to be undertaken.

Forest Carbon Partnership Facility (FCPF) This facility will have a capital fund of initially around US$300 million and it will focus its work on two mechanisms:

1 readiness (i.e capacity-building in up to 20 countries, focusing on developing a reference scenario and a REDD strategy, as well as creating an effective monitoring system with around US$100 to disburse);
2 carbon finance, which will work on demonstrational project transactions in around five countries with US$200 million to disburse.

The governance process is through a participants' committee and a broader participants' assembly.

Global Environment Facility (GEF)

GEF's climate change projects help developing countries and economies in transition to contribute to the overall objective of the United Nations Framework Convention on Climate Change and:

> *... to achieve ... stabilization of greenhouse gas concentrations in the atmosphere at a level that would prevent dangerous anthropogenic interference with the climate system. Such a level should be achieved within a timeframe sufficient to allow ecosystems to adapt naturally to climate change, to ensure that food production is not threatened and to enable economic development to proceed in a sustainable manner.* (UNFCCC, Article 2)

The GEF as the financial mechanism of the UNFCCC has around US$250 million a year to give to projects in the areas of energy efficiency, renewable energies and sustainable transportation. In particular, it manages two special funds under the UNFCCC — the Least Developed Countries Fund and the Special Climate Change Fund.

The GEF supports projects in:

- climate change mitigation: reducing or avoiding greenhouse gas emissions in the areas of renewable energy, energy efficiency and sustainable transport;
- climate change adaptation: aiming at increasing resilience to the

adverse impacts of climate change of vulnerable countries, sectors and communities.

The assembly of all GEF's members meets only every three years, so the main body for decision-making on an annual basis is the GEF Council. The council is made of 16 developing countries, 14 developed countries and 2 from the countries of Central and Eastern Europe and the former Soviet Union.

The decision-making is by consensus; but should there be a vote, then 'the council shall be taken by a double-weighted majority – that is, an affirmative vote representing both a 60 per cent majority of the total number of participants and a 60 per cent majority of the total contributions' (GEF, 1994). This is something that many developing countries are unhappy about.

There are questions on the future of the GEF. It will need to broaden its mandate and practices if it is to play a significant role in the future.

The Least Developed Countries Fund (LDCF) The LDCF was established to support a work programme to assist the least developed country parties to the convention. The focus of the fund is to help in the preparation and implementation of the National Adaptation Programmes of Action (NAPAs). The fund is administrated by the GEF.

Adaptation Fund (Stakeholder Forum for a Sustainable Future) There was a growing recognition of the need to also deal with adaptation measures. This fund was set up to finance adaptation projects and programmes in developing countries that are parties to the Kyoto Protocol.

The financing of the fund comes from proceeds to the Clean Development Mechanism. At present this is 2 per cent of the certified emission reduction (CER).

The governance structure for these funds is as follows: 16 members and 16 alternates. This consists of two representatives from each of the five United Nations regional groups and then one from the small island developing states; one from the least developed country parties; two others from the parties included in Annex I to the Convention; two others from the parties not included in Annex I to the convention (non-Annex I parties).

Special Climate Change Fund (SCCF) Established in 2001, the SCCF's remit under the convention is to finance projects relating to adaptation; technology transfer and capacity-building; energy, transport, industry, agriculture, forestry and waste management; and economic diversification.

The GEF was requested to take responsibility for creating the framework that the fund would operate under and then to manage the fund. The decision on this will be critical to who controls the funds for climate change in the future.

Clean Development Mechanism (CDM)

The Clean Development Mechanism allows developed Annex B countries with a greenhouse gas reduction commitment under the Kyoto Protocol to invest in projects that reduce emissions in developing countries. This is seen as a short- to medium-term opportunity to allow countries more time to reduce the emissions in their own countries.

There is some criticism of the mechanism, both in its operation and in its concept.

The CDM is overseen by the CDM Executive Board (CDM EB), which is under the guidance of the Conference of the Parties/Meeting of the Parties (COP/MOP) of the UNFCCC. CDM funds principally go to four countries: China, India, Brazil and the Republic of China, which account for around 80 per cent of the funds.

Joint implementation (JI)

Joint implementation is one of the three flexible mechanisms that was set up as a result of the Kyoto Protocol. It focuses on helping countries with binding greenhouse gas emission targets to meet their obligations. It focuses on countries wanting to reduce their emissions by investing in other Annex B country programmes to reduce emissions.

Some suggestions for the way forward

So what are the possible ways forward? The objective of this chapter is to give some idea of the complexity that exists in the world of multilateral climate change and energy within the UN system. The other chapters in this book clearly indicate the urgency that the world is facing as far as climate and energy is concerned. We believe that now is the right time to try and look at how the system might operate in the future. These suggestions are a contribution to that conversation.

At times, trying to quantify what is being undertaken in the name of climate change and energy within the UN system can seem overwhelming.

Perhaps the best place to start is a review of the mandates of UN agencies and programmes. This has been started through the Chief Executives Board and would benefit from a wider discussion, perhaps under the UN Economic and Social Council.

To help create a single UN operation, there should be a review of the mandates of each individual UN agency and programme in relation to climate change, against their present or future programmes of work. This could enable:

- the identification of overlaps with other UN bodies;
- the identification of gaps that need filling;
- the impact of the work being undertaken by the UN bodies to be assessed;
- additional funding needed to implement agreed objectives.

This could facilitate system-wide planning on climate change and energy to be undertaken. It would strengthen the role of the ECOSOC as a mechanism to deal with coordination between parts of the UN system.

In the area of the proliferation of different funds, a review of governance structures for the different funds could enable a learning process within the system as a whole. Many of these newer funds are trying different governance structures.

We are concerned that there needs to be a better link between the Bretton Woods institutions and UN ECOSOC, both in relation to guidance on system-wide activities on climate change, and in relation to the creation of a more coherent, transparent and accountable system for funding. Our suggestion would therefore be:

- a two-day high-level segment of ECOSOC dealing with system-wide coherence on climate change;
- an additional day of the Special High-Level Meeting of ECOSOC with the Bretton Woods institutions dealing with climate change funds.

One of the very good suggestions put forward by the High-Level Panel on System-Wide Coherence in 2006 was the setting up of a sustainable development board within ECOSOC. Such a board could deal with the reporting of the delivery at the country level of climate change activities through the One Country Programme approach. Although the idea had some institutional problems (such as what role there would be for UN-Habitat and UNEP), these are soluble. As such, a board would have real use in feeding back to ECOSOC on what is working and what is not.

The present state of discussions on sustainable development within the UN Commission on Sustainable Development and the UN General Assembly suggests that the present set-up isn't working. We believe there are a number of suggestions that should be considered that would enable a more coherent approach to sustainable development within the UN, and therefore in relation to climate change and other emerging and critical threats. These include:

- upgrading the UN Commission on Sustainable Development to a council of the UN General Assembly;
- creating a sustainable development security council;
- transforming the Trusteeship Council into an ecological security council.

The transformation of the CSD into a council of the General Assembly is following the lead of Human Rights, which did this in 2007. The idea of creating a sustainable development security council might go hand in hand with German Chancellor Angela Merkel's suggestion of creating an 'economic council' at the United Nations alongside the Security Council, and a UN charter on sustainable economics. If the chancellor's suggestion does go ahead, such a council could also become the overarching body for coordinating climate funds.

A new Earth Summit in 2012, as proposed by G77 and China, should tackle the issue of sustainable development governance. At the same time, it could deal with the issue of international environmental governance, which has ground to a halt after a number of years of work. The clustering of environmental conventions can be of great help in dealing with interlinkages.

If the Commission on Sustainable Development is going to continue, as opposed to becoming the preparatory process for a summit in 2012, then its multi-year programme could be revisited. It might want to address climate and energy issues as cross-cutting with the topics that the CSD is discussing.

A final word

We hope that this chapter fulfilled your appetite for burrowing into the UN system, and that the suggestions we have put forward are ones that give some food for thought.

As Albert Einstein said: 'We can't solve problems by using the same kind of thinking we used when we created them.'

Note

1 UN Energy aims to promote system-wide collaboration in the area of energy with a coherent and consistent approach since there is no single entity in the UN system that has primary responsibility for energy.

References

Brundtland, G. H. (1982) *Brundtland Commission: Agenda 21*, Oxford University Press, Oxford

Cering, D. (2000) *Outlook Weekly*, www.chinagate.cn/english/2283.htm

FAO (United Nations Food and Agriculture Organization) (2008) *FAO Mission Statement*, www.fao.org/climatechange/49358/en/

GEF (1994) Council Decision on Agenda item 5 GEF/C.2/CRP1 Geneva, Switzerland

Kennedy, R. (1968) Speech at City Club of Cleveland, Cleveland, OH, 5 April 1968, www.rfkmemorial.org/lifevision/onthemindlessmenaceofviolence/

Nussbaum, D. (2007) Speech at Stakeholder Forum London launch of UNEP's GEO4 Report, Stakeholder Forum, www.stakeholderforum.org/fileadmin/files/GEO-4_Launch_Report_-_Back_to_Our_Common_Future_December_2007.pdf

Steiner, A. (2007) *UNEP Global Environment Outlook 4*, Press Release, UNEP, www.unep.org/geo/geo4/media/media_briefs/Media_Briefs_GEO-4%20Global.pdf

UN (United Nations) (1992) *UN General Assembly Resolution on the Setting Up of the UN CSD A/RES/47/191*, www.un-documents.net/a47r191.htm

UN (2000) Chief Executives Board for Coordination of Higher Level Committee Programme, http://hlcp.unsystemceb.org/ Paris, France

UNEP (United Nations Environment Programme) (2007a) *Global Environmental Outlook 4*, www.unep.org/geo/geo4/media/

UNEP (2007b) *Nairobi Declaration*, www.unep.org/Documents.multilingual/Default.asp?DocumentID=287&ArticleID=1728&l=en

UNEP (2008) *Medium-Term Strategy 2010–2013: Environment for Development*, www.unep.org/PDF/FinalMTSGCSS-X-8.pdf

Vandeweerd, V. (2008) *Climate Change at the UNDP: Scaling Up to Meet the Challenge*, www.undp.org/climatechange/docs/English/UNDP_CC_Strategy.pdf

17

Who Decides? The Role of the United Nations and Security Council in Addressing Climate and Energy Insecurity

Chris Spence

Introduction

> *Our responsibility in this council is to maintain international peace and security, including the prevention of conflict. An unstable climate will exacerbate some of the core drivers of conflict such as migratory pressures and competition for resources. The recent Stern Report speaks of a potential economic disruption on the scale of the two World Wars and of the Great Depression. That alone will inevitably have an impact on all of our security, developed and developing countries alike.* (Margaret Beckett, UK foreign secretary, addressing the United Nations Security Council on 17 April 2007)

> *Climate change may have certain security implications, but generally speaking it is in essence an issue of sustainable development... Discussing climate change at the Security Council will not help countries in their mitigation efforts, nor will it help developing countries affected by climate change to respond more effectively to it.* (Liu Zhenmin, deputy permanent representative of China to the United Nations, speaking to the United Nations Security Council on 17 April 2007)

To ask whether the United Nations Security Council should take up climate and energy insecurity seems, on initial inspection, an almost pointless question. As other authors in this book have demonstrated, climate change and energy concerns present a major new security policy challenge. Climate change may be the most daunting problem of the century, posing a risk to international security and world peace. With a mandate to maintain international peace and security (Article 34 of the UN Charter)[1], it would seem only logical that the Security Council should play a significant role in addressing this challenge. Instead of asking why the UN Security Council should be involved in climate and energy concerns, it seems more reasonable to ask: why should it not? If this isn't an issue for the UN's most powerful body, what is? What other group could hope to tackle this global threat?

In fact, there are some valid objections to placing this matter in the hands of the Security Council – at least for now. Equally, there are compelling reasons for addressing the issue elsewhere.

This chapter considers the arguments both for and against placing the matter more firmly in the domain of the Security Council. It suggests that, when it comes to climate change and related energy issues, the primary focus of diplomats in the coming months and years should be elsewhere – namely, the ongoing talks taking place under the UN Framework Convention on Climate Change (UNFCCC) and its Kyoto Protocol. Talks under these treaties are at a critical stage, with negotiators seeking to reach an agreement by December 2009 on a global framework to tackle climate change after 2012. While climate and energy insecurity is not the primary focus of these talks, the UNFCCC process offers an unprecedented opportunity to set rules for how the international community mitigates and adapts to climate change through to 2020 and beyond. These rules could directly affect the pace and severity of climate change in the coming decades, which will, in turn, impact upon the level of security risk.

Finally, this chapter argues that although other processes and bodies certainly do have a role to play, they should in no way be allowed to jeopardize, supplant or distract from the current work under the UNFCCC and its protocol.

The role of the United Nations Security Council

On 17 April 2007, at its 5663rd meeting, the Security Council held an historic event – its first debate on the relationship between energy, security and climate (UN Security Council, 2007). The discussion was held at the request of the UK, which held the council presidency at that time.

The British wanted to address 'the security implications of a changing climate, including through its impact on potential drivers of conflict (such as access to energy, water, food and other scarce resources, population movements and border disputes)'. In a letter to the council, the UK pointed out – quite correctly – that no other major international forum had addressed these topics from this perspective. They also recognized that 'it is for other United Nations bodies (in particular, the United Nations Framework Convention on Climate Change) to pursue other aspects of climate change that are not within the mandate of the Security Council (including action to stabilize greenhouse gas concentrations in the atmosphere at a safe level, based on the principle of common but differentiated responsibilities and respective capabilities)'. In particular, the British sought a discussion on the potential impact of climate change upon key security risks, including border disputes, migration, energy supplies, other resource shortages, societal stress and humanitarian crises. Their pre-meeting letter on the subject presented a stark warning:

> *The cumulative impacts of climate change could exacerbate these drivers of conflict, and particularly increase the risk to those states already susceptible to conflict – for example, where weak governance and political processes cannot mediate successfully between competing interests.* (UN Security Council, 2007)

Other governments agreed with this assessment, and the debate was well attended. Many of the 55 speakers – from European neighbours to small island states in the Pacific – endorsed the UK's efforts to raise the profile of the issue and the potential role of the Security Council in addressing the security implications of climate change.

Objections and concerns

However, not all delegations were pleased. Among the dissenting voices was Ambassador Farukh Amil of Pakistan, who was speaking for the entire Group of 77 (G77) developing countries. Ambassador Amil argued that the Security Council was not the right venue for discussions on energy and climate change. These matters, he said, were better addressed under the UN General Assembly (UNGA), its Economic and Social Council (ECOSOC) and subsidiary bodies, including the Commission on Sustainable Development (CSD) and United Nations Environment Programme (UNEP). Furthermore, he noted that climate change was already addressed under a binding multilateral agreement – the UNFCCC. It was the Kyoto Protocol, not the council, which in his view was the proper forum to consider the risks associated with climate change. According to Ambassador Amil, the attempt to bring climate change and energy under the Security Council's umbrella was another example of the council encroaching on the authorities of other bodies, and compromising the rights of the UN's wider membership. In his opinion, 'no role was envisaged for the Security Council' on climate change (UN Department of Information, 2007).

The question for many developing countries was not whether climate change and energy insecurity were worth debating, but where they should be debated. For many in the Group of 77, the Security Council remains an exclusive bastion of power employed by the Council's permanent five member states to preserve and safeguard their own interests. Only one of the five permanent members, China, is a developing country. The other four permanent members – France,

Russia, the UK and the US – have little in common with the developing world. Many in the Group of 77 believe that it is better for developing countries if this issue is dealt with in other bodies – such as the General Assembly – where all nations have a voice:

> *For many developing countries, the decision by the UK presidency of the Security Council to hold a debate on climate and energy security held undertones of an inequitable response by the industrialized nations, such as the US and other global powers, most responsible for climate change.* (Dodds and Sherman, 2007)

Not all developing countries appeared to share this view. In 2008, the permanent missions of the Pacific small island developing states cosponsored a draft UN General Assembly resolution on the threat of climate change to international peace and security. The resolution noted that, if unaddressed, climate change will lead to security challenges and destroy small islands because of sea-level rise. It urged the Security Council to assess and mitigate this security threat (Climate-L.org News, 2008a).

Clearly, opinion is divided on whether the Security Council should have a role. However, any suggestion that the UK government introduced the debate into the council because it holds disproportionate power or control does seem a little far-fetched. In fact, the UK is widely regarded as having played a genuine leadership role on climate change among the Western powers, both domestically and internationally.[2] Statements by UK officials suggest that their aim was simply to use their presidency of a major international body to promote the importance of climate and energy issues at the highest possible level – just as they had in 2005 while holding the G8 presidency (G8, 2005). Margaret Beckett addressed this concern directly in her opening statement on 17 April 2007 when she said: 'We are not in this debate seeking to pre-empt the authority of those institutions and processes where action is being decided: the General Assembly, the Economic and Social Council and its subsidiary bodies and agencies, and, of course, the UNFCCC' (Beckett, 2007). Rather, her aim was to help inform and elevate the importance of these discussions.

While the motives of the UK seem to have been benign, the prevailing opinion among many developing countries of the Security Council has long been that it is among the most exclusive and least democratic bodies within the UN system. Developing countries feel the council has too many areas already under its sphere of influence. It is perceived as an elite and influential club limited only to a fortunate few. Whatever the UK's motivations, this negative perception undermined trust on the issue and the debate.

In recent years, there has been growing support for reforming the UN Security Council by broadening participation. While the current five permanent members can all mount strong cases for their continued inclusion, the global situation has clearly changed since 1945 when the Security Council came into existence. Other countries and regions are rapidly emerging and also want a seat at the table of power. While this chapter is not the place to examine such issues in detail, it is worth noting that until such reform occurs, accusations of exclusivity will continue. Broadening the council's membership might help to rebuild some of the confidence that has eroded in this institution[3]. In the context of climate change – which clearly requires a global solution – this is an important point. At this point in time, lack of trust could limit progress if climate and energy issues were pursued too vigorously in the council.

The role of the United Nations Climate Convention and other bodies

A second major argument against pursuing this issue in the UN Security Council is that there are other more relevant bodies in which to address climate change. In fact, climate change is currently being addressed at almost every level and in almost every body within the UN system. Whether it is under the United Nations Environment Programme, the United Nations Development Programme, the Commission on Sustainable Development, the Economic and Social Council, the United Nations Industrial Development Organization, or one of a dozen other agencies, climate change has become a key area of focus throughout the UN system. Since his arrival at the UN in January 2007, Secretary General Ban Ki-moon has given the issue a high priority. His focus is admirable.

Much of this work has been at the operational level among UN staff and consultants (Climate-L.org, 2008b). But, where should diplomats focus their attention?

The answer is clearly the UNFCCC and its Kyoto Protocol. Negotiations under the UNFCCC are at a critical stage. Diplomats are seeking to reach an agreement by December 2009 on a global framework to tackle climate change after 2012, which is the year when agreement on emissions limits for industrialized countries under the Kyoto Protocol expires.

While climate and energy insecurity is not the primary focus of these talks, the UNFCCC process has the potential to have a significant impact in this area, offering an unprecedented opportunity to set rules for how the international community both mitigates and adapts to climate change through to 2020 and beyond.

In December 2007, negotiators at the UN Climate Change Conference in Bali agreed to launch a comprehensive process to support sustainable implementation of the convention through long-term cooperative action. An agreement is to be reached by the 15th Conference of the Parties to the convention, which is taking place in Copenhagen in December 2009. The talks will seek to secure 'a long-term global goal for emissions reductions, to achieve the ultimate objective of the convention' (UNFCCC, 2007).

The ultimate objective of the convention, according to Article 2, is to achieve 'stabilization of greenhouse gas concentrations in the atmosphere at a level that would prevent dangerous anthropogenic interference with the climate system. Such a level should be achieved within a timeframe sufficient to allow ecosystems to adapt naturally to climate change, to ensure that food production is not threatened and to enable economic development to proceed in a sustainable manner'(UNFCCC, 2007).

Achieving the objective of the convention would therefore have a direct bearing on climate and energy security. Steps to achieve the convention's aim will have an impact upon the pace and degree of climate change in the coming decades. Such steps imply a dramatic change in any 'business-as-usual' scenarios envisaged for energy production and consumption in the coming decades. They will require serious political and financial commitments[4] (Stern et al, 2006).

At this time, the UNFCCC negotiations are the obvious forum in which to secure such an agreement. Arguably, they are the only forum (Frankels, 2007). Certainly, the UNFCCC process can boast broad support as the forum in which a post-2012 deal can and should be struck (Frankels, 2007).

Under the agreements reached in Bali, any Copenhagen deal will need to be more ambitious and comprehensive than commitments under the Kyoto Protocol. Since that treaty was inked, adaptation has joined mitigation as a critical focus, as has the need to address technology development and transfer and the need to address emissions from the destruction of forests. A future agreement will also need to include action both by developed and developing countries (Kyoto focused only on developed country commitments). As part of the agreement reached in Bali, developing country actions will be supported by industrialized countries in terms of technology transfer and financing. This message was reaffirmed more recently during the December 2008 round of negotiations in Poznań, Poland. Clearly, climate change is no longer viewed simply through an environmental lens. Under the UNFCCC talks, it is already recognized as a development issue and a serious economic concern. Long- and medium-term goals, pledges to limit greenhouse gas emissions and to control temperature rises and atmospheric concentrations of CO_2 – these are the outcomes that will build resilience and address climate and energy insecurity (Spence et al, 2008).

Whether these outcomes can be achieved in Copenhagen remains to be seen. Whatever happens in Copenhagen, though, the negotiating path will certainly not end there. But what is clear is that Copenhagen presents a genuine opportunity to make a dramatic breakthrough on climate change. A well-crafted agreement would provide a huge incentive for clean development and spur major changes in investment flows that could have an impact both upon the pace and extent of climate change, and upon the capacity of countries and communities to deal with the problem. These problems are at the heart of the climate and energy security debate (Spence et al, 2008).

Conclusions: Seeking security

There is no doubt that when it comes to climate change, the focus of the international diplomatic community should be on the UNFCCC. It is under this body that critical talks are taking place that will shape and direct future climate change and energy policy.

This is not to say that other processes and bodies do not have a role to play. Both within and outside the UN, regional and multilateral groups can contribute. The Major Economies Meetings – gatherings of key industrialized and industrializing countries – can play a part. So, too, can the Group of 8 during its annual gatherings. Arguably, both already have played a useful role[5] (UN Charter, Article 34). Even the Security Council could assist.

But these other processes should be secondary to the current work under the UNFCCC and its protocol. They should not distract from this central focus, and should be careful not to do anything that might erode trust at this critical period on the road to Copenhagen. In the coming months and years, it is the diplomatic efforts under the UNFCCC and Kyoto Protocol that represent the international community's best hope of forging a global agreement on the scale needed to combat global warming and limit future climate and energy insecurity.

Notes

1 Under Article 34 of the UN Charter, the Security Council may 'investigate any dispute, or any situation which might lead to international friction or give rise to a dispute, in order to determine whether the continuance of the dispute or situation is likely to endanger the maintenance of international peace and security'.

2 The UK has taken on commitments under the Kyoto Protocol and announced numerous measures at the domestic, regional and international level.

3 Many books, articles and newspaper columns have been devoted to the issue of Security Council reform. For an overview of the issue, see Wikipedia, 'Reform of the UN Security Council', www.en.wikipedia.org/wiki/Reform_of_the_United_Nations_Security_Council.

4 The need for significant action on mitigation and adaptation, including significant increases in investment patterns and new technologies, have been widely noted and endorsed by many groups. For instance, leaders of 16 of the world's largest economies recognized this at a summit in Hokkaido, Japan, in 2008; see www.iisd.ca/recent/recentmeetings.asp?id=5#mtg5587.

5 As an example of the contributions that these two processes have attempted to make towards the post-2012 negotiations under the UNFCCC, see two articles in the 11 July 2008 issue of *Linkages Update*: 'G8 Summit endorses long time climate change vision' and 'Hokkaido Toyako Major Economies Meeting adopts declaration, minus targets', www.iisd.ca/whats_new/whatsnew110.html.

References

Beckett, M, (2007) Speech to the UN Security Council, 17 April 2007, Webcast of the Security Council Meeting on the Relationship between Energy, Security and Climate, 17 April 2007, webcast.un.org/ramgen/sc/sc070417am.rm

Climate-L.org News (2008a) *Draft General Assembly Resolution on Threats to Climate Change to International Peace and Security to be Discussed*, International Institute for Sustainable Development, New York, 12 August 2008, www.climate-l.org/2008/08/draft-general-a.html

Climate-L.org News (2008b) International Institute for Sustainable Development, New York, www.climate-l.org/

Dodds, F. and R. Sherman (2007) 'Climate and the UN: A new bid for control?', *BBC Online*, 18 May 2007, www.news.bbc.co.uk/2/hi/science/nature/6665205.stm

Frankels, F. (2007) 'Formulas for quantum emission targets', in J. E. Aldy and R. N. Stavins (eds) *Architectures for Agreement: Addressing Global Climate Change in the Post Kyoto World*, Cambridge University Press, Cambridge, UK

G8 (2005) *Chairs Summary*, July 2005, www.g8.gov.uk/servlet/Front?pagename=OpenMarket/Xcelerate/ShowPage&c=Page&cid=1119518698846

Spence, C., K. Kulovesi, M. Gutiérrez and M. Muñoz (2008) 'Great expectations: Understanding Bali and the climate change negotiations process', *Review of European Community and International Environmental Law*, vol 17, no 2, pp142–153

Stern, N., S. Peters, V. Bakhshi, A. Bowen, C. Cameron, S. Catovsky, D. Crane, S. Cruickshank, S. Dietz, N. Edmonson, S.-L. Garbett, L. Hamid, G. Hoffman, D. Ingram, B. Jones, N. Patmore, H. Radcliffe, R. Sathiyarajah, M. Stock, C. Taylor, T. Vernon, H. Wanjie and D. Zenghelis (2006) *Stern Review: The Economics of Climate Change*, HM Treasury, London

UN Department of Information (2007) *Report/Media Statement: Security Council Holds First Ever Debate on Impact of Climate Change on Peace, Security, Hearing Over 50 Speakers*, 17 April 2007, www.un.org/News/Press/docs/2007/sc9000.doc.htm

UNFCCC (2007) *Bali Action Plan*, (Decision 1/CP13, document FCCC/CP/2007/6/Add.1, www.unfccc.int/resource/docs/2007/cop13/eng/06a01.pdf#page=3

UN Security Council (2007) 'Letter dated 5 April 2007 from the Permanent Representative of the United Kingdom of Great Britain and Northern Ireland to the United Nations addressed to the President of the Security Council', Document S/2007/186, www.un.org/Docs/journal/asp/ws.asp?m=S/2007/186

Zhenmin, L. (2007) Webcast of the Security Council Meeting on the Relationship between Energy, Security and Climate, 17 April 2007, webcast.un.org/ramgen/sc/sc070417am.rm

18

Financing the Low-Carbon Transition in Developing Countries: The Role of Multilateral Development Banks

Barbara J. Bramble

Introduction

A wide range of national and international financial institutions have important roles to play in providing the necessary resources to help developing countries provide modern energy supplies to all their citizens using low-carbon fuels.[1] Dangerous climate change cannot be avoided without accomplishing both energy access for all and moving away from fossil fuels. This chapter analyses the role of the multilateral development banks (MDBs)[2] in financing the transition to a low-carbon future because they have the dual role of promoting economic development to address poverty, as well as providing a portion of the needed finance for energy. While the total amount of energy-related funding contributed by these institutions to emerging economies has been dwarfed in recent years by private commercial investors, the MDBs remain critical lenders to poor countries. They also influence private-sector finance because of the 'seal of approval' their participation brings to projects that they invest in, and their policy advice remains influential.

Our conclusion is that the MDBs are not yet playing the role they need to play in order to be part of the solution. Most are still lending to major greenhouse gas-emitting industries, such as fossil fuel and mineral extraction, coal-based power and livestock production. Most of them have new programmes or policies to encourage energy efficiency and renewable energy supplies, and most are slowly increasing their investments in these areas; but these sectors do not loom large in their lending portfolios. On the other hand, at the World

Bank there is much experimentation in carbon trading, and multiple new partnerships and 'funding mechanisms' have been created. This could be seen as a 'razzle dazzle approach', which diverts attention from the difficult chore of reducing the greenhouse gas intensity of the industries supported through World Bank Group's mainstream lending and policy operations. Because these banks are the creatures of their governing boards (principally the donor nations), one should conclude that this situation reflects the desires of the countries that govern the institutions.

The challenges of the 21st century: Achieve the Millennium Development Goals and move to solve climate change

Much attention is at last being paid to the severe challenge that climate change and its associated impacts will pose for all nations, particularly the developing countries. The *Stern Review: The Economics of Climate Change* and other studies have made it clear that while the poorer nations have done the least to cause global warming, they will suffer the earliest and most severe consequences (Stern et al, 2006). Access to modern sources of energy is at the base of all initiatives to realize the Millennium Development Goals (MDGs), which are aimed at significantly reducing poverty by 2015. So financing the energy transition in developing countries has the dual challenge of contributing to the solutions for global warming[3] while at the same time providing more high-quality energy to the poor to safeguard their development aspirations.

The threat of global warming is now recognized as so serious that all international financial institutions must turn their attention to this priority. The major private-sector banks and other investors all have the opportunity to direct their lending priorities towards global warming solutions, as they make investment decisions. But financial institutions that have access to public funds or guarantees[4] have an additional obligation to act with regard to the consequences for the future of the planet and to benefit the poorest. This applies to official development assistance funds, clearly, but also the export credit agencies, as well as the MDBs. The question for this chapter is: what should be the appropriate role for the World Bank and the other MDBs in helping the world to slow down climate change and move towards a sustainable energy future?

Very brief notes on the long history of controversy over multilateral development banks (MDBs) and energy

Since at least the mid 1980s, civil society groups in many countries have pushed the MDBs, particularly the World Bank, to change their approach to energy lending. Environmental and social advocates have asked that the emphasis on massive central grid projects be shifted to smaller-scale decentralized projects that could be built on an appropriate scale and schedule to be more affordable, and provide energy access to rural poor communities, often far from the grid. Studies had shown, for example, that massive hydroelectric dams and power plants not only took many years to plan and build, but also caused massive relocations of local populations, often without fair compensation, and the resulting power was often exported far away to cities, thus making millions of poor people even poorer.[5] After the UNFCCC was signed at the Rio Earth Summit in 1992, the drumbeat for reforms in the MDBs' energy lending grew stronger, with demands that it end subsidies for fossil fuels and other environmentally damaging energy production, consult widely with local communities, require fair resettlement plans, pursue energy efficiency and pollution reduction, and generally pay attention to the climate change and sustainable development consequences of energy projects. By 1993, the World Bank agreed to several of these demands in new policies to be applied to energy lending and advice to client governments.[6]

But the new energy policies were honoured mostly in the breach, and in 1996 were downgraded to 'good practice' guidelines. Several internal studies recommended reforms but were ignored. In 1998 a report by the Secretariat of the Global Environment Facility concluded that the World Bank had not 'taken meaningful action to reduce its traditional role as financier of fossil fuel power development' (GEF, 1998).

Controversy over MDB-funded energy projects in numerous countries continued to grow throughout the past decade. The World Bank first embraced and then rejected the findings of an independent body called the World Commission on Dams, established to study the social, economic and environmental impacts of large hydroelectric dam projects (World Commission on Dams, 2000). It then launched the extensive three-year Extractive Industries Review (EIR) to study impacts upon the poor of oil and mining projects, which produced a significant body of recommendations based on wide consultation with civil society and business, as well as governments. The findings of the EIR included that the Bank should 'phase out investments in oil production by 2008 and devote its scarce resources to investments in renewable energy, resource development, emissions reducing projects, clean energy technology, energy efficiency and conservation, and other efforts that delink energy use from greenhouse gas emissions' (Extractive Industries Review, 2003). The Bank rejected most of the recommendations, but did commit to increasing lending for renewable energy by 20 per cent each year for five years.[7]

Meanwhile, the financial role of the MDBs was evolving too. From the 1990s on, private-sector commercial banks, export credit agencies such as the UK Export Credit Guarantee Department and the US Overseas Private Investment Corporation, and sovereign wealth funds of countries such as Singapore, China and Brazil increased the amount of capital investment in emerging economies. The explosion of foreign direct investment (FDI) in the last 15 years, in at least a select list of countries, has dwarfed the amount that the MDBs provide.[8]

Now, while the poorest countries still do not have access to commercial foreign investment funds, the MDBs can still be significant lenders. But in the emerging economies and middle-income developing countries, the MDBs find themselves in the odd position of competing for the most 'bankable' projects with commercial lenders. Here the MDBs' policy advice and the imprimatur they provide to projects that they approve are often more influential than the funds they can lend.

The MDBs have long been criticized for investing in the expansion of industrial sectors that significantly ratchet up greenhouse gas emissions, without any methodology in place to predict or monitor the consequences or to consider the real costs and benefits of alternatives to either the borrowing countries or the globe. Major projects have been financed to promote fossil fuel extraction, aluminium smelting and agriculture, all primarily for export, that have modest impacts upon host country development or improving local employment levels. So the borrowing countries have been saddled with industries of the past, and with their attendant environmental and social impacts, but relatively little economic development to ease their transition to better livelihoods in a low-carbon future.

Current programmes and operations of selected MDBs

World Bank Group

The controversies outlined above formed the back drop for the 2005 G8 Summit at Gleneagles, Scotland, where the World Bank was charged by the G8 leaders with taking 'a leadership role in creating a new framework for clean energy and development'. In this ambitious global plan, the World Bank outlined what is needed to provide energy for development, to accelerate the transition to a low-carbon economy and to adapt to climate change. It included a range of activities by governments and financial institutions of all kinds, including the MDBs. This Clean Energy Investment Framework (CEIF) immediately came under fire for failing to concentrate on providing access to energy by the poorest, and for its emphasis, once again, on centralized government-dominated programmes, including 'clean coal', large-scale hydropower and nuclear-fired power plants.[9]

And, indeed, the National Wildlife Federation's examination of the World Bank Group's energy lending portfolios in 2001 and 2007 bears this out.[10] We found that the overall amount devoted to fossil fuel development and power production soared from US$433 million in 2001 to almost US$2.1 billion in 2007. And the investments in large-scale hydro grew from US$77 million to US$361 million.[11]

The World Bank's International Development Association (IDA) sought replenishment funds from its participating donors at the end of 2007. One of the key reports submitted to make the case for new funding emphasized the need to promote both greenhouse gas reductions and clean energy for the poor. But it set no benchmarks or targets and made no commitments for action. In fact, it misleadingly stated that 'projects, such as "clean" energy investments, may yield lower net benefits compared with "traditional" energy projects' (World Bank, 2007, p18).

This statement reveals that the Bank primarily considered a limited range of financial costs, but neglected the long-term costs and the health and energy access benefits for the poor of many clean energy alternatives. Given that there are still 1.6 billion people without access to electricity, (UNDP, 2007, Figure 1.7), one could argue instead that *most* of IDA's energy funding should be allocated to solutions that are appropriate for poor populations, who often reside far from energy grids – for example, small-scale hydro, solar and wind – that are based on locally available resources and that support domestic consumption. Instead, in 2001 and 2007, the overall combined energy lending amount for the IDA and the International Bank for Reconstruction and Development (IBRD) stayed essentially the same, while the proportion devoted to fossil fuel development went up and the proportion for renewables and efficiency went down.[12]

The International Finance Corporation (IFC) has recently expanded its overall capital available for private-sector loans, and its energy portfolio has grown commensurately. As the size of its lending portfolio has grown substantially, the absolute amount of finance devoted to renewables and efficiency has also grown significantly, from US$12 million in 2001 to over US$175 billion in 2007. But the proportion devoted to these purposes has hardly moved.[13] Instead, the IFC is funding more and larger greenhouse gas-emitting industries. Its energy portfolio and the pipeline of projects under review are dominated by oil, gas and coal, and also by large hydro, which often is a source of methane emissions.

The World Bank claims to be fulfilling its promised 20 per cent per year increase in lending for energy efficiency and renewables. But in the latest CEIF progress report, only 3 out of the 12 projects listed as the renewable energy portfolio, and about half of the efficiency investments, are actually theirs. All the others are carbon trades or Global Environment Facility (GEF) projects.

Moreover, while the World Bank group had imposed a tacit moratorium on the funding of large coal-fired power plants in the 1990s, this has ended. In 2008, the IFC board approved a new 4000MW (five units of 800MW each) coal-based power plant, to be built in Gujarat State, India, called Tata Ultra Mega. Other such projects are in the approval pipeline.

Aside from energy production, many other sectors contribute to global warming. Deforestation and agriculture are among the most significant sources of greenhouse gas emissions, especially in developing countries. Here, too, the IFC's loans have invited criticism. Unlike the rest of the World Bank, which has refrained from financing of large-scale livestock operations for years, the IFC has dived right in. The Bank's cautious approach to livestock investments was based on analysis of the significant environmental and social impacts of commercial feedlot systems and industrial milk, poultry and pork production, as well as the negative implications of such projects for food security (World Bank, 2001). Moreover, livestock production is now considered to be one of the major causes of greenhouse gas emissions, not only because of methane resulting from the animal's digestion, but also because in certain locations they stimulate deforestation for grazing land and feed production. Since 2001, the IFC has invested approximately US$750 million in livestock projects totalling US$2.22 billion. These include the recently approved Bertin Cattle Ranching

Project in Brazil, approved in 2007, which local experts consider will increase the incentives for deforestation and potentially threaten a number of indigenous peoples' reserves in the area.[14] The internal assessment of this project by the Bank's Independent Evaluation Group stated that 'the project poses a grave risk to the environment and to the Bank's reputation', specifically referring to the risk from this one project of deforestation on a scale 'roughly equivalent to Jamaica'.[15]

Similarly, in 2006, the IFC also provided finance for Indonesia's Wilmar Oil Palm project without considering any of the potential risks of deforestation, or the resulting greenhouse gas emissions and impacts upon indigenous peoples and biodiversity.[16, 17]

But instead of taking on the hard work of shifting the priorities of the lending and policy operations of their traditional portfolios, the World Bank has recently chosen to shift attention to a suite of new climate change financing mechanisms.[18] It is almost as if they are applying for a new job: manager of all the tools for channelling funds of other donors and the private sector into carbon reductions.

All of these new funds and related efforts are expanding knowledge and experience about carbon markets, and although they are often criticized by civil society groups, they seem to be worthwhile initiatives.[19] But it is important for the World Bank not to confuse these innovative mechanisms with their basic job of promoting sustainable low-carbon development among poor nations. These initiatives should not be diverting attention from the need to transform their own overall fossil fuel-intensive lending plans, and their policy advice and capacity-building work, which should be promoting energy access for the poor through clean renewable technologies.

The Inter-American Development Bank (IDB)

The IDB has a similar chequered past, in terms of funding projects that are accused of stimulating deforestation, such as the old BR 364 Highway into the Brazilian Amazon, and the more recent Camisea gas pipeline in Peru. But their energy lending portfolio has been tending away from oil drilling and other fossil fuel development for years,[20] and their investments in large hydro projects have dropped from US$206 million in 2001 to less than US$100 million in 2007. On the other hand, our analysis of their loans showed almost no renewable energy or small-scale hydro projects in 2001 or 2007. Instead, they are facilitating the technical assessment of scores of proposed infrastructure projects in South America under the rubric of the initiative for Infrastructure Integration for the Region of South America (IIRSA). This conglomeration of up to 400 waterways, ports, dams, roads and reservoirs has been called a 'giga-project' by observers in the region because of its unprecedented breadth and scope. The potential for forest destruction in the Amazon and other areas as a consequence of the synergistic and cumulative impacts of all these major works, and the related agriculture production that they are aimed to stimulate have not been fully evaluated. But observers expect massive new greenhouse gas emissions and regional rainfall pattern changes as a result.

As a new direction, the IDB recently launched a Sustainable Energy and Climate Change Initiative (SECCI) that is based on four 'pillars':

1 promoting renewable energy and efficiency;
2 biofuels development;
3 increasing access of its borrowers to carbon finance; and
4 adaptation to climate change.

While biofuels production can be a source of additional greenhouse gas emissions if land-use change is stimulated,[21] the other pillars are essential and positive new directions for the IDB and we will follow their development with interest. But it is impossible at this juncture to tell if this initiative will be significant enough to actually influence the overall lending portfolio. The real question for the institution, just as noted above for the World Bank, is not whether another finance mechanism is added to the mix of new carbon 'gizmos', but whether the overall lending decisions and policy advice in the sectors of energy, agriculture, transportation and others actually assist countries to move away from traditional greenhouse gas intensive development. Significantly, the SECCI doesn't yet propose lending targets or have any other goals for influencing the mainstream operations of the IDB to ease the transition to a low carbon future.

The Asian Development Bank (ADB)

The ADB has been in the midst of a review of its energy strategy, as well as its environmental and social safeguards, for the last couple of years, seeking to articulate its most effective role in the region for the medium-term future. Civil society groups have criticized the new draft safeguards for being vague and unenforceable. But at least in rhetoric they contain welcome new requirements: that the borrowers will 'promote the reduction of ... greenhouse gases'; quantify direct and indirect emissions from the project; evaluate options to reduce emissions; and pursue appropriate options.[22] In addition, logging in 'primary moist tropical forests or old growth forests' is not eligible for ADB lending.

Moreover, the latest draft of the new energy strategy squarely states that ADB operations in the energy sector should be realigned to 'meeting energy security and transition to a low carbon economy' (ADB, 2008). Recommendations include expanding their energy efficiency investments; support for renewable sources of electricity; technical assistance for the needed regulatory reforms; as well as assisting borrowers to gain access to the Kyoto Protocol's Clean Development Mechanism (CDM) revenue to cover additional costs of renewable energy installations. And the ADB strategy recommends continuing to avoid funding oil and gas exploratory projects. However, the strategy states that coal 'will remain a major source of energy for electricity and heat' and therefore the ADB will 'actively promote' its extraction and use (ADB, 2007, p24).

The European Bank for Reconstruction and Development (EBRD)

The EBRD, like the World Bank, has already started a carbon trading fund as manager of a US$35 million fund from The Netherlands. Otherwise, its energy portfolio is rather mixed. Using the same categories as described above for our analysis of the IDB and the World Bank, the National Wildlife Federation's snapshot of available information indicates that the EBRD's fossil fuel lending in 2007 was double (US$845 million) that in 2001 (US$466 million). But since their overall energy lending had also doubled in that period, the fossil fuel proportion remained at the same level. Meanwhile, our analysis shows its investments in renewable energy, efficiency and small hydro to have grown significantly, from US$24 million in 2001 to over US$300 million in 2007. Moreover, it avoided large hydro entirely in those years.

Pressure for change at two export credit agencies (ECAs)

In 2007, the ECAs of both Europe and the US pledged and/or were asked to make big changes in their lending operations with respect to climate change.[23] On 29 November 2007, the European Parliament passed a resolution calling for the 'discontinuation of public support, via export credit agencies and public investment banks, for fossil fuel projects' (European Parliament, 2007).This vote will still need implementing legislation, but it's a clear call for reform.

The Overseas Private Investment Corporation (OPIC) is the principal ECA of the US. It has been under pressure for years to recognize the impacts of its financial operations on a number of areas, such as on human rights, indigenous peoples and the environment, particularly global warming. Even though it has been prohibited from providing direct investment loans to finance oil and gas extraction, in 2005 alone OPIC provided US$580 million of other support to oil and gas projects, including pipelines. Therefore, in 2007 legislation was proposed to reduce the greenhouse gas emissions associated with OPIC projects, and to require OPIC to give preference to renewable energy and energy efficiency investments. While this was being considered by the Senate Foreign Relations Committee,[24] OPIC announced an initiative 'to reduce greenhouse gas emissions in OPIC-supported projects by 20 per cent in the next ten years and shift emphasis to renewable and energy efficient projects'.[25]

As a group, however, the ECAs have not yet incorporated climate change within their investment decisions. Ironically, several of the largest private-sector commercial banks have taken the initiative to adopt principles for fossil-fuel lending, which include encouraging energy efficiency and other low-carbon alternatives, and evaluating investment risks in the context of future carbon regulation.[26]

Positive new directions to promote the transition

Since experts, civil society organizations and advocates have been pushing for a move away from fossil fuel-based projects for decades, what confidence should we have that another set of such recommendations would have any more effect this time around?

On the one hand, one might say: not much. The 'pressure to lend' that has been in place since at least Robert McNamara's term as World Bank president has caused staff and board to measure job performance by the quantity of funding moved to the borrowers. This has meant that modest renewable energy and efficiency investments of appropriate scale are rarely viable in the race to place loans in amounts of hundreds of millions of dollars.[27] In speeches, papers and personal communications, staff at all levels of the MDBs have long revealed a fixation with the need to lend money 'at the speed of business' and to become as efficient in making loans as the private sector in order to compete – to win a share of the loan business of developing countries. For example, new World Bank President Robert Zoellick's first public speech emphasized how the Bank 'would be simplifying the procedures and cutting interest rates... We aim to be faster, better and cheaper.'[28] But they have it exactly wrong. Instead, this chapter argues:

- Where there is competition for making these loans, then the MDBs are not needed. Why should public money be provided where the private sector is active?
- Their role is to promote development and poverty reduction. That is not the primary purpose of the private-sector banks. They also operate in countries that are not yet attractive to foreign direct investment. So they are in a different business from the private sector institutions.
- The private banks want the imprimatur of the MDBs, the seal of approval – so in that sense the MDBs are still indispensable. But this will only be true if they really do help projects to succeed in human development and environmental terms.
- The only way for the MDBs to add real value to the private-sector business is to undertake thorough assessment of social and environmental impacts, to understand how to put development front and centre and to provide for social and human capital needs; this often takes attention and time.

So if they insist on giving up their main value added, the MDBs will be putting themselves out of business – especially since they are now required, as a result of public pressure, to assess social and environmental impacts before making decisions. Thus, they can never be the fastest out of the blocks. This fundamental misreading of their role may be beyond repair, and if so they will be slowly rendered irrelevant in a race with the private sector that they can't win. But if they focus on their role as public-sector funders, the following recommendations, which are neither original nor new, may yet be heeded.

Fundamental reforms

Energy access The role of public-sector financial institutions should be to concentrate on providing clean energy to the poor, especially those without access to electricity. This can be accomplished through lending operations as well as policy advice.[29] Energy investments and policy work should be dedicated to demonstrating the development and anti-poverty benefits of clean energy and efficiency, and to paving the way for private-sector investments in these sectors.

Analysis of costs and benefits The MDBs, along with their borrowers, must undertake rigorous assessments of the costs and benefits of alternative investments, based on full-cost accounting. Renewable energy and efficiency improvements can often be more cost-effective, quicker to build, less prone to cause corruption and conflict, while creating more local jobs and reducing social and environmental impacts.

These assessments must include greenhouse gas accounting and 'shadow pricing'. Because of the impacts upon the poor of global warming, all decisions on MDB operations, including financing as well as policy work, should be accompanied by an analysis of the greenhouse gas implications of alternative courses of action. This estimate can be part of the normal environmental and social assessment process, and must include both direct and indirect emissions, for the entire life of any project (without comprehensive greenhouse gas accounting of their operations, the MDBs and their boards have been able to ignore their long-term impacts to accelerate global warming).

'Shadow pricing' enables decision-makers to compare alternatives over time, using a reasonable estimate of future carbon prices to assess potential costs or revenues. The MDBs are in an excellent position to work with the GEF, the Clean Development Mechanism and other sources of grant funding to buy down any additional upfront costs of low-carbon alternatives before carbon pricing becomes routine.

Global warming screen Based on the greenhouse gas accounting, the MDBs should put a 'screen' on all energy-intensive investments, just as they now do with ozone-depleting chemicals production and child labour, to prohibit certain classes of projects and use extra care with others. In this way they can help their borrowers to reduce greenhouse gas emissions as they increase access to energy for the poor.

The MDBs can be especially influential by working with the financial intermediaries they support through on-lending operations to use similar greenhouse gas emissions screens.

Reduction targets Aggressive targets should be set for progressive reductions of the greenhouse gas emissions of the existing loan portfolios and projects in the pipeline. This includes both energy production and energy intensive industrial and agricultural development.

Above all, the boards of directors of the institutions, and especially the donor countries, must face up to their responsibility for approving loan portfolios that increase rather than decrease greenhouse gas emissions.

Sectors that need special precautions or moratoria

The MDBs should follow the recommendations of the Extractive Industries Review. No new lending should be made to fossil fuel development such as coal mining, tar sands or oil and gas production; these extractive industries have a sorry record of corruption and failure to address the energy needs of the poor or to promote solid development of local economies.

Large hydropower (above 10MW), especially in tropical areas, should not be considered a renewable or 'clean' technology – heed the advice of the World Commission on Dams.

Forestry and deforestation are the largest sources of greenhouse gas emissions in many developing countries. The MDBs must stop funding deforestation via livestock, agriculture and industrial forest concessions. These projects have a record of failing to help the poor, on top of the greenhouse gas impacts. Specifically, they should prohibit any funding for projects that fragment or convert natural forests for any purpose.

They should put a comprehensive moratorium in place to end investments in commercial-scale livestock and related feed production and infrastructure. Focus on healthy food, mixed farming for smallholders and local markets.

The MDBs should confront 'carbon capture and storage' technology with the same scepticism with which they have long treated renewable energy.

Finally, MDBs' role in carbon trading needs scrutiny because of real or perceived conflicts of interest, the MDBs should confine their role to developing the models, and then phase out their participation. They should not be long-term fund managers.

Notes

1 There is a wide range of topics related to the responsibility of developed nations (and the financial institutions they support) for providing measurable, reportable and verifiable financing to developing nations for climate change adaptation, mitigation, reducing deforestation and technology transfer, whether under the United Nations Framework Convention on Climate Change (UNFCCC) or other agreements. This chapter looks only at the specific question of the role of international financial institutions in the low-carbon transition.

2 This term is used here to refer to financial institutions that are owned and managed by multiple countries, such as the World Bank, the Inter–American Development Bank, the Asian Development Bank, the African Development Bank and the European Bank for Reconstruction and Development, including all of their private-sector and insurance arms.

3 While it is clear that the industrialized countries are primarily responsible for historic greenhouse gas emissions, and have the obligation to reduce their emissions under the UNFCCC, in order to avoid dangerous climate change all nations will eventually have to cooperate to reduce overall emissions on the basis of common but differentiated responsibilities.

4 While some finance channels for the MDBs involve direct donations, most of their capital has been in the form of bonds issued on the basis of 'callable capital', which is guaranteed by the member governments.

5 Besides the specific MDB energy policies and project decisions, a more general critique by experts and coalitions of non-governmental organizations (NGOs) from countries in both the North and the South for many years has documented the lack of disclosure of essential information for informed decisions, failure to ensure appropriate consultations with affected populations, and lack of accountability for negative social, health and environmental impacts upon women, indigenous peoples and other vulnerable groups.

6 See the World Bank (1993a, 1993b). During this period, most of the MDBs improved their policies on a wide range of needed reforms, from transparency and participation of local communities in project evaluations and implementation, to establishment of independent inspection panels or ombudsmen to hear complaints about failure to implement the improved policies. Unfortunately, these reforms have not been sufficient to remedy two fundamental problems:
   1 the imbalance in the governing structure of the MDBs (where voting power on the boards is allocated according to 'one dollar one vote');
   2 what is called the 'pressure to lend', which to this day underlies the MDBs' relentless search for bankable projects and 'high-risk, high-reward' ventures, such as large power dams and fossil fuel development (see Rich, 1994, for an analysis of the pressure to lend and a history of the civil society MDB reform campaigns).

7 Draft World Bank Management Response to the Extractive Industries Review, World Bank, 4 June 2004. The Bank claims to reach this target each year, but only by counting GEF projects and carbon trading projects – which civil society groups argue is playing with other people's money.

8 In 2006, foreign direct investment (FDI) in developing countries topped US$325 billion, and even official development assistance was up over US$100 billion, while the proportion provided by the five major MDBs has remained static at around US$50 billion. Among the developing countries that have received the bulk of FDI are China, Hong Kong, Brazil, Mexico, India and Singapore (International Monetary Fund, 2006).

9 Sustainable Energy Economy Network (2006, pp15–20).

10 The World Bank Group comprises several parts, including the International Bank for Reconstruction and Development (IBRD), the original part of the Bank; the International Development Association (IDA), the 'soft loan window' for the poorest countries; and the International Finance Corporation (IFC), which lends to the private sector.

11 Our staff reviewed the project data on the Bank's website and allocated projects to categories, such as 'fossil fuels', 'renewables, efficiency and small hydro', 'large hydro' and 'administrative improvement and repair', based on analysis of available information.

12 Using the categories described in note 11, and the figures for the IBRD and IDA, fossil fuel projects totalled US$139.5 million in 2001 and US$243.5 million in 2007, while projects classified as renewables, efficiency and small hydro came to US$165 in 2001 but only US$93.5 million in 2007. See also Bank Information Centre (2008).

13 Our assessment of the public records reveals that the IFC's energy portfolio soared from about US$330 million in 2001 to US$2.375 billion in 2007. Meanwhile, the proportion devoted to projects identifiable as efficiency investments and renewables rose only from a negligible 3.6 to 7.5 per cent.

14 See Amigos da Terra Amazonia Brasiliera (2008).

15 Letter from senior adviser, Independent Evaluation Group, World Bank, www.wellfedworld.org/worldbank.htm.

16 Project no 25532, approved in 2006. The IFC labelled this project as a Category C, meaning that it would have 'minimal or no direct, adverse social or environmental impacts'.

17 In addition, the IDA/IBRD interventions in the forestry sector, in Cambodia and the Congo, are expected to open whole new areas to logging, which may become significant new sources of greenhouse gas emissions.

18 In 2000 the World Bank launched the Prototype Carbon Fund with 6 governments and 17 industry partners to serve as an early model of how markets can be used for promoting greenhouse gas reductions. It is a project-based fund with a US$180 million portfolio of 24 projects. In 2003 the US$128 million Community Development Carbon Fund opened, with 9 governments and 16 private partners, to invest in greenhouse gas reduction projects, many in Africa, that provide 'community development benefits'. The Bio Carbon Fund was established in 2004 to demonstrate cost-effective greenhouse gas reductions through carbon sequestration in forests or agro-ecosystems. Its first tranche raised US$53.8 million, and a second tranche opened in March 2007. In addition, the World Bank manages about a dozen other funds in cooperation with specific donor nations such as Italy, Spain and The Netherlands to develop carbon reduction projects that may qualify under the Kyoto Protocol's Clean Development Mechanism (CDM) and Joint Implementation (JI) programme. It is putting together an Umbrella Carbon Facility to aggregate funds from many public and private participants to purchase emission reduction credits associated with large CDM and JI projects.

19 In 2008 the Bank's board approved two new facilities: the Clean Technology Fund to use additional donor funding to provide relatively discounted finance for low-carbon technologies in the power, transportation, construction and agriculture sectors; and the Strategic Climate Fund aimed at highly vulnerable countries. Also, in 2008, the Bank launched a Forest Carbon Partnership Facility to assist 20 or more countries to develop capacity and test several models of using carbon markets to promote the reduction of emissions from deforestation and degradation (REDD). This will involve both a readiness fund to help countries assess their forest stocks and pilot deforestation reduction policies, as well as a carbon fund to purchase emissions reduction credits.

20 For example, some groups have pointed out that the World Bank has influence on the rules of the CDM, and has actively lobbied for a more 'investor-friendly' definition of 'additionality', while it profits from deals as manager and trader for several of the carbon funds. Moreover, its lending portfolio has been supporting fossil fuel projects with lifetime greenhouse gas emissions many times bigger than the amount of emission reductions anticipated under the CDM. In addition, some of the biggest CDM trades involve HFC-23 destruction projects, which manufacture credits at little or no cost, while garnering premium payments. See Wara (2006).

21 A recent report in the Dow Jones Newswires revealed, however, that the IDB is likely to provide up to US$1.5 billion to the state-owned oil company Petro–Ecuador: see www.bicusa.org/en/Article.3535.aspx, 14 October 2007.

22 See the development of the Roundtable on Sustainable Biofuels, www.epfl.ch and www.bioenergywiki.net, for further information on the need for sustainability standards, including avoiding both direct and indirect land-use change, to ensure that biofuels don't create more greenhouse gas than they save.

23 ADB (2007, p27). The National Wildlife Federation (NWF) snapshot of the ADB's

lending portfolio in 2001 and 2007, using the criteria described in note 11, shows that fossil fuel and hydro lending went up; but their investments in renewables and the category we called 'admin/repair' rose more dramatically, consistent with the draft strategy.

24 ECAs provide government-backed loans, guarantees and insurance to corporations of their home countries to expand business opportunities in developing countries or other emerging markets, and most industrialized countries operate such funds.

25 Press release, www.opic.gov/news/pressreleases/2007, 14 June 2007.

26 See Carbon Principles signed by JP Morgan Chase, Morgan Stanley and Bank of America in 2008: www.carbonprinciples.com.

27 There is a long history of the World Bank looking for its fundamental role in development finance, and it has lurched from one doctrine to another over the decades. See Rich (1994) for an analysis of this evolution. Since the period after World War II, when the Marshall Plan took over their first job of reconstructing Europe, the World Bank has moved through phases from large infrastructure investments to structural adjustment programme loans, and then to a period in the1990s when they turned to social and environmental issues, lending for education and other 'softer' approaches to development under President James Wolfensohn. But now the harder approach is back: it's all about winning the race with the private sector.

28 See MacDonald (2007).

29 Experts at the ADB have suggested that the 'ADB's greatest contribution' in the energy sector may be through policy dialogue and advice to build the necessary policy, legal and regulatory frameworks to promote energy efficiency, renewable energy and cleaner technologies. See ADB (2000).

References

ADB (Asian Development Bank) (2000) *Special Evaluation Study on Energy Policy 2000 Review: Energy Efficiency for a Better Future*, Operations Evaluation Department, Asian Development Bank, Manila, Executive Summary, pvii

ADB (2007) *Draft Energy Strategy*, May 2007, p24, ADB, Manila

ADB (2008) *Draft Safeguard Policy Statement*, ADB, Manila, October, pp43–44

Amigos da Terra Amazonia Brasiliera (2008) *The Cattle Realm: A New Phase in the Livestock Colonization of Brazilian Amazon*, Amigos da Terra Amazonia Brasiliera, Brazil, p4, www.amigosdaterra.org.br; www.amazonia.org.br/guia/detalhes.cfm?id=236645&tipo=6&cat_id=46&subcat_id=198

Bank Information Centre (2008) 'World Bank's lending for fossil fuel skyrockets as it positions itself as the "climate bank"', Bank Information Centre, www.bicusa.org/en/Article.3940.aspx, 23 July

European Parliament (2007) *Trade and Climate Change*, www.europarl.europa.eu/sides/getDoc.do?Type=TA&Reference=P6-TA-2007-0576&language=EN

Extractive Industries Review (2003) *Striking a Better Balance: The World Bank Group and Extractive Industries*, December, World Bank, Washington, DC

GEF (Global Environment Facility) (1998) *Study of the GEF's Overall Performance*, 2 March, GEF, Washington, DC

MacDonald, L. (2007) 'Global development: Views from the center', http://blogs.cgdev.org/globaldevelopment/2007, 10 October 2007

Rich, B. (1994) *Mortgaging the Earth: The World Bank, Environmental Impoverishment, and the Crisis of Development*, Beacon Press, Boston, MA

Stern, N., S. Peters, V. Bakhshi, A. Bowen, C. Cameron, S. Catovsky, D. Crane, S. Cruickshank, S. Dietz, N. Edmonson, S.-L. Garbett, L. Hamid, G. Hoffman, D. Ingram, B. Jones, N. Patmore, H. Radcliffe, R. Sathiyarajah, M. Stock, C. Taylor, T. Vernon, H. Wanjie and D. Zenghelis (2006). *Stern Review: The Economics of Climate Change*, HM Treasury, London

Sustainable Energy Economy Network (2006) 'How the World Bank's energy framework sells the climate and poor people short,' pp15–20. See www.seen.org published by Bank Information Center, Bretton Woods Project, Campagna per la Reforma Della Banca Mondiale, CEE Bankwatch Network, Friends of the Earth International, Institute for Policy Studies, International Rivers Network, Oil Change International and Urgewald, 18 September

UNDP (United Nations Development Programme) (2007) *Human Development Report*, UNDP, New York, NY, Figure 1.7

Wara, M. (2006) *Measuring the Clean Development Mechanism's Performance and Potential*, Working Paper no 56, July, Program on Energy and Sustainable Development, Center for Environmental Science and Policy, Stanford University, US

World Bank (1993a) *The World Bank's Role in the Electric Power Sector: Policies for Effective Institutional, Regulatory and Financial Reform*, World Bank, Washington, DC

World Bank (1993b) *Energy Efficiency and Conservation in the Developing World*, World Bank, Washington, DC

World Bank (2001) *Livestock Development: Implications for Rural Poverty and Environment and Global Food Security* Directions in Development, World Bank, Washington, DC

World Bank (2007) *IDA and Climate Change: Making Climate Action Work for Development*, IDA Sustainable Development Network, Washington, DC, 11 October, World Bank, Washington, DC, p18

World Commission on Dams (2000) *Dams and Development: A New Framework for Decision Making*, Earthscan, London

19
European Union Response to the Challenge

Stavros Dimas[1]

One of the first policy documents that I presented to the European Union (EU) Commission after I was appointed as the commissioner for environment was called *Winning the Battle against Climate Change*. This policy document, or *Communication*, as we call it in Brussels, was one of the most important achievements of the Barroso Commission because it outlined the key elements for the EU's climate change strategy. It became the blueprint for the European response to climate change and, with the support of the member states (particularly the UK presidency), the main ideas were taken up by the international community at the 2005 Climate Conference in Montreal. However, there was one thing that we got wrong, and that was the title. As the scientific evidence accumulates, it is clear that the fight against climate change is much more than a battle. It is a world revolution that we need to engage in with great determination if we are serious about the climate crisis we are facing.

Damaged economies, refugees, political instability, international security and the loss of life are typically the results of war; but they will also be the results of climate change if we allow it to continue unchecked. Our fight against climate change can be likened to a drastic change because in order to reduce emissions something very like a revolution in the economy is needed. All sectors – transport, energy, agriculture and foreign policy – must work closely together to meet a common objective. This is also a sign of interdependence because every country in the world will be affected by the impacts of climate change and should contribute to fighting it.

In today's world of accelerating globalization there are many challenges that nations simply cannot address working on their own: fighting terrorism, pandemics such as bird flu, poverty in Africa, nuclear proliferation, financial and economic crisis, energy security and climate change. These are all areas where countries have to find a way of working together if their policies are to have any chance of success. Indeed, climate change presents the most urgent, pressing and serious challenge to an increasingly globalized world and there is no alternative to global cooperation. No country can tackle climate change in isolation. The UK's emissions of greenhouse gases are responsible for approximately 2 per cent of the annual global total. EU emissions are responsible for 14 per cent. The contribution of all is needed and a global response is the only possible solution. There are a number of reasons why this will need a strong and effective European Union, both to contribute to the fight against climate change and to maintain our international leadership.

First and foremost, effective 'climate diplomacy' is essential if we are to convince the US and key developing countries to participate constructively in negotiation. The EU has over 450 million citizens and has the largest market in the world. It is self-evident that by acting together we have a much greater diplomatic influence than by acting alone.

Second, the EU must lead by example. We have already made progress with regard to emission reductions; the latest EU inventory of greenhouse gas (GHG) emissions for 2006 shows that total EU-27 greenhouse gas emissions were 10.8 per cent below base year levels without emissions and removals by land use, land-use change and forestry and were 0.3 per cent lower compared to 2005. At the same time the EU-27 economy grew by 3 per cent in 2006 compared to 2005. Total GHG emissions in the EU-15 were 2.7 per cent below the emissions of the Kyoto base year. Since 1990, the EU–15 economy (expressed as gross domestic product, or GDP) grew by almost 40 per cent. The respective projections show that – with current efforts and additional measures – the EU-15 will reduce their emissions by 11.3 per cent and over-achieve their Kyoto Protocol emission reduction commitment of an 8 per cent reduction. EU-27 will reduce their emissions by 16.3 per cent compared to base year levels.

However, while it is very encouraging that we have already made huge progress in cutting emissions as the European economy grows, it is clear that many member states need to accelerate their efforts to limit emissions further, notably after 2012. With our Climate and Energy Package agreed last year, the EU has now developed the necessary tools that will enable it to reduce its GHG emissions by at least 20 per cent by 2020 compared to 1990.

Third, the EU has developed the world's first example of a cross-border Emissions Trading Scheme (ETS) to reduce greenhouse emissions. It is the world's first and largest international emissions trading scheme and covers some 11,500 installations in the EU, accounting for almost half of EU carbon dioxide emissions. Set up in January 2005, we are still in the early stages of the scheme; however, early academic research indicates that emissions have fallen by several per cent compared with their levels before the start of the EU ETS. All indicators suggest that the ETS is one of the most cost-effective policy tools that exists for reducing emissions, and it is a model that the rest of the world is already looking to as we work towards a global approach to emissions control.

The fourth motivation for a strong and united EU is that key EU policies also need to be mobilized against climate change. These include the obvious sectors such as transport and energy. Research policy and industrial policy are also needed in order to develop and promote new technologies. Inside the EU, regional funds and agricultural policy will have to support investments in emissions reduction and climate change adaptation. Development cooperation will have to support similar investments in developing world countries, and the EU's agricultural policy will notably have to protect the soils and forests that absorb carbon dioxide.

The fifth and final reason relates to competitiveness. Industry often claims that high environmental standards could damage competitiveness; but a common European approach applies across the entire single market. It therefore creates a level playing field in the EU while allowing the environmental objective to be met.

All together this constitutes a convincing case for a strong and effective EU stance regarding climate change. Let's now turn to international action: getting an ambitious and comprehensive climate agreement in Copenhagen is now our priority since EU efforts alone will not be sufficient to limit climate change.

The science on climate change is now clear and the impacts of today's climate change striking. In the summer of 2007, the North-West Passage shipping route through the Arctic became navigable for the first time since records began. The area covered by Arctic sea ice has shrunk to the smallest ever seen. The situation in the Arctic Sea is only one of many signs that climate change is happening.

We have a good idea of the likely social, environmental and economic impacts of both determined action but also of no action, based on the Intergovernmental Panel on Climate Change (IPCC) and Stern reports. New technologies need to be developed; but it is first and foremost essential to ensure a speedy deployment of existing low-carbon and energy-efficient technologies. For this to materialize, we need to get the right incentives and regulatory environments in place as quickly as possible. This is all the more so since we have the resources to make the necessary investments. Our greatest challenges are not scientific or technical or even economic. The challenges are political. At present we have political commitments from Europe's leaders to lead the fight against climate change; but when it comes to making difficult economic choices, there is still a great gap between rhetoric and reality.

The EU Commission's starting point in relation to the science is that climate change must be limited to no more than 2°C above pre-industrial temperatures. The scientific evidence indicates that the risks of irreversible and potentially catastrophic impacts will greatly increase beyond this threshold. Science tells us that if we are to have even a 50 per cent chance of keeping within this 2°C limit, worldwide greenhouse gas emissions will have to peak before 2020 and then fall by at least 50 per cent of 1990 levels by 2050. As the next step towards this longer-term reduction, and as part of a new global climate agreement, the commission is proposing that developed countries cut their emissions to an average of 30 per cent below 1990 levels by 2020 and that developing countries reduce their emissions by 15 to 30 per cent below business as usual by 2020.

An international agreement is a necessary condition for meeting our objective of limiting temperature increases to below 2°C. Irrespective of this, the EU is in any event committed to a reduction of its emissions by at least 20 per cent by 2020.

There are a number of reasons for taking such an independent commitment. First, this strengthens our leadership in the international context by showing determination for action. Second, it will benefit our economy in terms of increased energy security and public health. Third, it gives a welcome signal to the market that the emissions trading system will continue after 2012. Last, but not least, it enhances predictability and encourages investment in clean technologies.

A firm commitment by the EU and the rest of the developed world to tough emissions reductions is essential if we are to convince developing and transition countries to take action. But without the cooperation of these countries, it will simply not be possible to keep global warming within 2°C. By 2020 they will be emitting more than developed nations.

Turning to the developing world, the rise in emissions from developing countries, and in particular from emerging ones, needs to start slowing down as soon as possible; from around 2020, their emissions need to be reduced in absolute terms. This will be a major challenge; but one can be optimistic: there are many options for cutting emissions that would deliver immediate economic and social benefits and that would not affect their legitimate pursuit of economic growth and poverty reduction. To take one example, major potential exists in energy efficiency that should undoubtedly be the focus of short-term investments. This will obviously require upfront investment, but will generate a net benefit in the mid term. I am confident that developing countries will engage in this with the required determination. In fact, this is already happening: China, India, Brazil, South Africa and Korea already have, or are in the process of developing, climate and energy plans. This should be further encouraged and supported, bilaterally and multilaterally.

Another challenging issue will be to tackle deforestation. We cannot continue to overlook this since it is the second largest emissions source after fossil fuel combustion, and contributes around 20 per cent of global emissions. It is also one of the main reasons for the loss of global biodiversity. Our figures indicate that for the 2°C ceiling to be met, net deforestation needs to be halted completely within the next two decades and then reversed through reforestation and afforestation schemes. This is now part of the climate negotiations where I hope we will come to an agreement on an incentive mechanism, rewarding performance-based policies supported by adequate finance.

These proposals are ambitious; however, the bottom line is that they are essential if we are to keep global warming within manageable limits and spare future generations the most devastating economic, social and environmental impacts of climate change. As the Stern Review[2] has underlined, the benefits of taking action far outweigh the costs. Effective international action on climate change means going beyond national self-interest and pooling sovereignty.

Clear and new scientific evidence since the last IPCC report produced in 2007 all points in the same direction: climate change is happening and accelerating at a faster rate than previously predicted. This calls for urgent action and an international agreement to be reached in Copenhagen at the end of 2009 is imperative.

There is no excuse for delaying action: we need to show determination and be creative in devising an ambitious and fair international framework. As the EU demonstrated with the adoption of its Climate and Energy Package in a record time, when political will is there, success is around the corner.

Notes

1 This chapter is based on a number of press releases and statements on behalf of Stavros Dimas, including Stavros Dimas (2007) *EUROPA Press Release: Climate Change: Why a Global Response Needs European Leadership*, Launch event of the European Commission and the All Party Group on Climate Change Cooperation for 2007, http://europa.eu/rapid/pressReleasesAction.do?reference=SPEECH/07/8&format=HTML&aged=1&language=EN&guiLanguage=enarliamentary; Stavros Dimas (2005) *EUROPA Press Release: Climate change: Commissioner Dimas Welcomes 2005 Reduction in EU Greenhouse Gas Emissions and Calls for Further Action*, http://europa.eu/rapid/pressReleasesAction.do?reference=IP/07/835); Stavros Dimas (2006) *EU News: Giving Kyoto a Future*, http://jpn.cec.eu.int/home/news_en_newsobj1587.php#.

2 See Stern, N., S. Peters, V. Bakhshi, A. Bowen, C. Cameron, S. Catovsky, D. Crane, S. Cruickshank, S. Dietz, N. Edmonson, S.-L. Garbett, L. Hamid, G. Hoffman, D. Ingram, B. Jones, N. Patmore, H. Radcliffe, R. Sathiyarajah, M. Stock, C. Taylor, T. Vernon, H. Wanjie and D. Zenghelis (2006) *Stern Review: The Economics of Climate Change*, HM Treasury, London.

20
Sub-National Climate Change Action Model for the World

Linda Adams and Steve Howard

Introduction: The importance of state and regional action

Governments at all levels recognize the urgency of acting on climate change. People in regions around the world already find themselves adapting to a changing planet. Science points to the responsibility of governments to act now in order to avoid the worst impacts predicted by future climate change scenarios.

We also have a better understanding of the costs of not taking immediate steps to mitigate climate change. According to the latest update of the Stern Review, the costs of taking action, now estimated at approximately 2 per cent of gross domestic product (GDP), are far outweighed by the costs associated with adopting a 'wait and see' attitude. The cost of not acting on climate change could amount to the equivalent of losing at least 5 per cent of global GDP per year.

For years, state and regional governments around the world have demonstrated tremendous leadership on tackling climate change. States, provinces and regions are on the frontline when it comes to dealing with the impacts of climate change. They prepare for and deal with the aftermath of extreme forest fires, storms, flooding and drought. These events affect their budgets directly and the decisions they make must therefore account for the threatening changes that come with higher temperatures. Many have begun implementing programmes and strategies to reduce greenhouse gas (GHG) emissions and develop plans to address adaptation needs.

Sub-national governments are well placed to pioneer the opportunities presented by the transition to a low-carbon economy, having greater flexibility in setting and implementing policy. State, regional and provincial governments also have the advantage of being able to take into account the specific characteristics, vulnerabilities and strengths of their citizens and regions. Moreover, states and regions have responsibilities that differ from the national level. They are often responsible for environment, resource and energy policy and the use of public lands, as well as public transport, waste management and access to water. They establish and/or implement codes and standards for buildings, appliances and vehicles. They set taxation and subsidy policies for business and agriculture and they have a variety of regulatory and infrastructure arrangements with municipal governments. This means that the decisions taken by states and regions can actually shift the marketplace towards low-carbon solutions such as energy efficiency, sun, wind and geothermal power – and, in the case of emission standards – new vehicle technologies such as plug-in hybrid and electric vehicles.

National governments can, in turn, take advantage of the experience gained at the state and regional level. A recent example of a dramatic shift in the right direction is President Obama's review of the US Environmental Protection Agency's (EPA's) earlier denial of California and other states' right to raise their vehicle emission standards. This is just one example of the important work that states and regions are undertaking at the sub-national level, which, with national partnership, creates world class emission reduction standards and charts a course for a new low-carbon economy.

Climate change is increasingly gaining centre stage at international meetings such as the G8. National governments are playing host to symposiums focused on strengthening partnerships and identifying climate change solutions. And progress, although slow, is being made at the annual United Nations climate change conferences where the core aspects of an international climate change agreement are debated.

The 2007 United Nations Climate Change Conference in Bali, Indonesia (COP 13), marked an important juncture in the international negotiations. Parties to the United Nations Framework Convention on Climate Change (UNFCCC) negotiated how to address climate change in the post-2012 period when the Kyoto Protocol's first commitment period ends. Bali culminated in the adoption of the Bali Road Map and Action Plan, which laid the building blocks for future negotiations. Governments agreed to slow the growth of emissions and eventually cut them; provide financial support for climate change mitigation; cooperate on measures for adaptation; and work on reducing deforestation and technology transfer. Negotiations in Bali, and more recent negotiations in Poznań, Poland, are all important steps toward the COP 15 conference in Copenhagen in 2009. This is the timeline that governments have set to come to a global agreement for the post-2012 period.

We must remember that any new international agreement will not take effect until 2012. In the interim, state and regional governments are contributing to shorter-term progress by diligently putting into practice climate change policies and programmes, such as building efficiency standards, automobile emission standards, low-carbon fuel standards, renewable portfolio standards (policies that require a proportion of an area's energy generation to be from renewable sources), and sustainable land-use policies. Sub-national governments are already implementing reduction strategies ahead of the next global agreement by sharing best practices and new technologies across developed and developing country boundaries.

Origins of sub-national cooperation: From the Montreal Declaration to today

States and regions have been acting on climate change for many years. The process was energized by the 2005 Climate Leaders Summit in Montreal, held in conjunction with the 11th Conference of the Parties (COP 11) to the United Nations Framework Convention on Climate Change. Sponsored by the Climate Group and the provincial governments of Quebec and Manitoba, this meeting brought together sub-national government and business leaders from around the world to discuss how they could further action on climate change, both individually and collaboratively.

The summit resulted in the Montreal Declaration, which has been signed by more than 30 sub-national governments from across the US, Canada, Europe, Africa, Australia and South America. The declaration commits signatories to setting short- and long-term GHG emission reduction goals. A range of measures to achieve these objectives, including energy efficiency, renewable energy, clean transport, fiscal mechanisms and sustainable forestry and agriculture programmes, are laid out in the declaration,. What's more, it commits signatories to work together, across continents, to speed up the shift to a global low-carbon economy.

Since the 2005 Climate Leaders Summit, leaders from states and regions from around the world have come together to work with the Climate Group through a Climate Alliance to progress the commitments in the Montreal Declaration. Today the 38 regions gathered in the Climate Alliance account for approximately 3 per cent of global population, 12 per cent of global GDP and 5.6 per cent of global carbon emissions.

Sub-national action: Then and now

Since signing the Montreal Declaration, sub-national governments throughout the world have met and exceeded the commitments outlined in the document. States and regions have expanded partnerships in China and India. They have also become test beds for new low-carbon technologies.

In California, we have seen both the urgency of dealing with climate change and the opportunities that low-carbon policies represent at the sub-national level. Governor Schwarzenegger set aggressive greenhouse gas reduction targets in 2005 to reduce emissions to 1990 levels by 2020. In 2008, he signed those targets into law through Assembly Bill 32, the Global Warming Solutions Act of 2006. Assembly Bill 32 establishes a comprehensive programme to achieve real, quantifiable and cost-effective reductions in global warming pollution.

California is already implementing an array of programmes and initiatives to ensure that the targets are met. One example is the California Solar Initiative, a ten-year US$2.9 billion programme announced in January 2006. This initiative provides rebates to make solar more affordable to consumers. The goal is to increase the amount of installed solar capacity on rooftops in the state by 3000MW by 2017. Moreover, California's Low Carbon Fuel Standard (LCFS), issued by Executive Order in 2007, calls for a 10 per cent reduction in the carbon content of the state's transportation fuels. The standard is expected to reduce greenhouse gases by 15 million metric tonnes in 2020 and decrease dependency upon petroleum.

In July 2008, California redoubled its energy savings efforts by establishing new targets for the years 2012 through 2020 for regulated utilities. Within investor-owned utility (IOU) service areas, these goals are expected to save over 4500MW: the equivalent of over 9 major power plants, and over 16,000GWh of electricity savings and 620 million therms. Combined with recent goal setting by California's publicly owned utilities under Assembly Bill 2021, these goals provide an aggressive contribution to state-wide energy savings targeted under Assembly Bill 32.

And it doesn't stop there. On 18 September 2008, the California Public Utilities Commission adopted the California Long-Term Energy Efficiency Strategic Plan, which includes four 'big bold strategies' as cornerstones for significant energy savings with widespread benefit for all Californians. All new residential construction must be zero net energy by 2020, and commercial construction by 2030. The heating, ventilating and air-conditioning (HVAC) industry will be reshaped to deliver maximum performance HVAC systems. Finally, the California EPA has initiated a Low-Income Energy Efficiency (LIEE) programme in which all eligible low-income customers will be provided cost-effective energy-efficiency measures in their residences by 2020.

Many other sub-national governments around the world have taken innovative steps to reduce greenhouse gas emissions. Colleagues in Canada are amongst those leading the way. British Columbia, for example, a mountainous province on the west coast of Canada, is the first North American jurisdiction to introduce a revenue neutral carbon tax. All carbon tax revenue will be recycled to British Columbians through tax reductions. British Columbia also requires that all new electricity projects have zero net GHG emissions and is the first state in North America to legislate the requirement for the public sector to be carbon neutral by 2010. Manitoba, a prairie province in the heart of Canada, has been harnessing geothermal technology. As a result of innovative incentives and loan programmes, Manitoba is the North American leader in ground source heat pump installations, quadrupling sales in recent years.

Ontario, the most populated province in Canada, is undertaking a coal phase-out initiative – shutting down coal-fired plants by 2014 – which will reduce GHGs by up to 30 million tonnes. Ontario is also protecting at least 225,000 square kilometres of the Far North Boreal Region, giving priority protection to key ecological features such as endangered species habitat. Ontario and Manitoba are also partnering to establish and protect an inter-provincial wilderness area covering more than 94,000 square kilometres. Canada's French-speaking province of Quebec has implemented a duty on gasoline and fossil fuels that will ensure Cdn$200million (US$168 million) per year to fund actions to meet its target of reducing emissions to 6 per cent below 1990 levels by 2012. Moreover, Quebec is targeting a 300,000 tonne reduction of carbon dioxide equivalent (CO_2e) from the agricultural sector through Cdn$24 million (US$20 million) in funding for technologies such as biogas capture and processing at liquid manure installations.

Regional governments in Europe are also moving quickly to address this pressing issue. Brittany, a large peninsula in the north-west of France, aims to generate 1000MW of energy from land-based wind farms in 2010 and an additional 500MW from offshore turbines by 2015. The Basque Region in Spain is developing a renewable portfolio standard that will deliver 3.75GWh annually by 2012, and in Germany, Bavaria has set the ambitious goal of doubling its primary energy consumption from renewables – from 8 to 16 per cent – by 2020.

The autonomous community of Catalonia, located in Spain on the eastern coast of the Iberian Peninsula, is promoting cleaner transportation policies. Through its CO_2 per kilometre-based car matriculation tax, people are provided with the incentive to buy cleaner, less polluting vehicles. A similar approach has been taken in Flanders, Belgium, where the regional government will now be able to collect car registration taxes and annual road tax for passenger cars, both having been adjusted so they promote the purchase of environmentally friendly cars.

North Rhine Westphalia, the westernmost and most populated federal state in Germany, is undertaking an initiative called 'North Rhine Westphalia saves energy'. This aims to achieve a 20 per cent reduction in overall primary energy consumption across the economy by 2020 compared to 2006. The government encourages energy refurbishment of residential and social-sector buildings through consultancy-type services to citizens and businesses as well as vast investment programmes. For example, 130 million Euros (US$165 million) will be injected in order to renew the energy infrastructure of schools and day-care facilities for children.

Scotland has set some ambitious targets that include a goal of 50 per cent total renewable energy generation by 2020 to become a zero waste society and to increase forest cover by 25 per cent by 2050. Moreover, with the aim of improving energy performance and investing in renewable energy, the Wallonia government, Belgium, is granting subsidies to local authorities to improve the performance of public lighting and of municipal, provincial and regional buildings such as schools and hospitals. At the same time, all Welsh Assembly government-influenced projects are required to meet the Building Research Establishment Environmental Assessment Method (BREEAM) Code for Sustainable Homes Level 4. The Welsh Assembly also aims for all new buildings in Wales to be zero carbon from 2011.

In Australia, states and regions are also jump-starting climate change policy. South Australia will soon be home to the largest single rooftop photovoltaic installation in Australia with 10,000 square metres of panels saving around 1400 tonnes of GHGs each year. Victoria's energy efficiency efforts are set to reduce GHG emissions by 8.1 million tonnes and deliver annual average household energy savings of 45 Australian dollars (US$30). Also, since 2004 all new homes built in Victoria have been required to reach a 5-star energy rating standard.

Right here in the US, colleagues also are actively taking up the challenge. For example, in 2007 energy-efficient investments in Connecticut resulted in a US$4 return on every US$1 invested and lifetime savings of 4.3 billion kilowatt hours and 2.6 million tonnes of CO_2. And in 2008, New York announced its goal of cutting electricity usage by 15 per cent by 2015 through its Energy Efficiency Portfolio Standard (EEPS).

As briefly mentioned above, the California Clean Car standard is an excellent example of a regulatory measure that will significantly reduce greenhouse gas emissions in California and the 14 other states that have adopted the same standards. The standard requires a 30 per cent reduction in greenhouse gas emissions from cars and light trucks by 2017, which makes up nearly 20 per cent of the overall reduction goal in California alone. Just one measure that targets the transportation sector can make a huge dent in the problem.

Significant progress in reducing GHG emissions has also been made at the sub-national level in China and India and other emerging economies. Guangdong, the richest and second most populated province in China, has been dealing with energy pollution and severe energy shortages over the past decade. As a result, the province is placing great emphasis on improving energy efficiency and is already the least energy-intensive province in China. In India, Maharashtra, the largest industrial state, is delivering energy saving benefits through its Energy Conservation Plan that far outweigh the costs of implementation.

In São Paulo, Brazil's most populated state, a number of Clean Development Mechanism (CDM) projects have been developed, including biogas to energy initiatives in landfills, breweries and farms. The Bandeirantes Landfill, for example, generates 20MW from the scheme and saves about 1 million tonnes CO_2e per year. Over on another continent, the province of Western Cape, located in the south-west of South Africa, is working to ensure that renewable energy technologies such as solar heating are being mainstreamed within all public buildings.

Unfortunately, there are still states and regions that have yet to commit to acting on climate change. What these sub-national governments need to take into account, apart from the ethical implications of inaction, is that they will be at a competitive disadvantage economically if they maintain a business-as-usual or 'wait and see' attitude. The economy is shifting under our feet, and these states and regions are not taking the opportunity to properly position themselves to take advantage of the opportunities of a low-carbon economy.

While there is still much work to do, the momentum from other sub-national governments around the world who have embraced the need to act now is truly inspiring and hopeful.

From sub-national to international cooperation on climate change

The efforts of sub-national governments are significant not only on their own, but because we have agreed to work together across borders and across oceans. In addition to leading on climate actions at home, sub-national governments have formed partnerships with each other as a model for international cooperation on climate change. Out of the Montreal Declaration and state and regional partnerships, key bilateral agreements have been signed. For example, California has signed a memorandum of understanding (MOU) with South Australia, Manitoba, British Columbia and many others to advance areas such as climate change legislation, the expansion of solar and geothermal power and the adoption of tailpipe emissions standards.

Not only are states and regions advancing bilateral partnerships, but they are also advancing expansive North American-wide initiatives to tackle climate change. Three regional cap-and-trade initiatives are currently being established in North America, spanning the East, the West and the Midwest. The first such initiative that arose is the Regional Greenhouse Gas Initiative (RGGI or 'ReGGie'). Created in 2005, this is the first mandatory, market-based effort in the US to reduce greenhouse gas emissions. Membership includes ten north-eastern and Mid-Atlantic US states, while Eastern Canadian provinces are participating as observers. This initiative is targeting a 10 per cent carbon dioxide (CO_2) emissions reduction from the power sector by 2018.

On the West Coast, the Western Climate Initiative (WCI) emerged in 2007. This regional initiative has partners and observers from the US, Canada and Mexico. Partners currently include Arizona, California, New Mexico, Oregon, Washington, Manitoba, British Columbia, Utah, Montana, Quebec and Ontario. The states and provinces have the goal of reducing GHG emissions 15 per cent below 2005 levels by 2020 in different sectors across the economy through market-based mechanisms such as cap and trade.

The Midwest also jumped on board in 2007 with a Midwestern Greenhouse Gas Reduction Accord that will develop multi-sector, market-based mechanisms to achieve reduction targets, including cap and trade. The Midwest has also agreed to actively collaborate on such issues as energy efficiency and renewable energy.

An important mechanism in North America is the Climate Registry. This is a collaborative partnership among US, Canadian and Mexican states, provinces and tribes to develop and manage a common reporting system for GHG emissions. If we want to be effective in the fight against climate change, we need mechanisms that help us to measure, track, verify and publicly report GHG emissions, and the Climate Registry has been developed to do just that.

From California's perspective, we would like to see a federal cap-and-trade programme alongside complementary policies built upon, and compatible with, state and regional action. Success in combating climate change will require action at both the national and state levels. States have long been the incubators of innovation and led on policies such as energy efficiency, clean transportation, renewable energy, low-carbon fuels and many more.

Sub-national action doesn't stop there. States and regions have been building momentum towards effective international networks, working towards faster and more effective climate change action. In Bali in 2007, sub-national governments reaffirmed their engagement towards the objectives of the Montreal Declaration and told many of their inspiring success stories. Moreover, 80 regional participants from around the world recently met at the 2008 St Malo World Summit of Regions, which was hosted by the President of Brittany and organized by the United Nations Development Programme (UNDP) and the Network of Regions for Sustainable Development (NRG4SD). The participants committed to partnerships between developed and developing country regions in areas such as the development of climate plans, low-carbon technology and best practice transfer.

California was proud to host the Governors' Global Climate Summit in 2008, which brought together regions from all over the world, including China, India, the US, Canada, Mexico, Brazil and Indonesia. Participants committed to focusing efforts on the largest emitting sectors, including forestry, cement, iron, aluminium, energy and transportation. States and regions also agreed to employ strategies to fight climate change such as technology transfer, incentive programmes, sharing of best practices and market- or non-market-based programmes. What is exciting about this type of climate change action is that it is a win–win situation – that is, these strategies can be implemented almost immediately and at little or no cost with substantial economic and environmental benefits. Governments also committed to focusing research, development and deployment activities on areas such as energy efficiency, renewable energy and zero and low-carbon electricity generation and fuels.

United Nations support for state and regional action also is increasing. The UNDP, the United Nations Environment Programme (UNEP), the Climate Group and other networks of regions are moving forward on setting up the UN–Regions partnership Towards Low Carbon Emission and Climate Change Resilient Territories. The aim, over a period of five years, is to:

- Ensure at least 500 regions from developing and emerging countries are trained on the post-Kyoto process, the carbon footprint for emission diagnosis and mapping vulnerability for adaptation diagnosis, the existing and forthcoming carbon finance mechanisms and available technologies.
- Accompany at least 50 regions from developing countries in assessing their emission and vulnerability diagnosis and elaborating their mitigation and adaptation strategy and action plan.
- Design a whole set of public policy and investments projects in these 50 regions to be submitted to the different financial opportunities such as official development assistance (ODA), public–private partnerships (PPPs), decentralized cooperation, and the carbon finance mechanisms such as the Clean Development Mechanism (CDM), the Adaptation Fund, the Global Environment Facility (GEF), UN-REDD (reducing emissions from deforestation in developing countries) and the Millennium Development Goals Carbon Facility.

The global carbon market is already reaching billions of US dollars and this is projected to increase in a post-Kyoto regime. Yet, these funds are not being adequately accessed by developing countries due to a lack of sound bankable projects. Reasons for this include a lack of well-defined targets at the national level, limited knowledge of the opportunities available and poor know-how on the design of such projects.

The UN–Regions partnership is investing in additional human resources and aims to fundraise an estimated US$1 million per region to support developing and emerging regions in the establishment of a climate change strategy and action plan, and to generate projects in each region. The final target is to reach more than US$45 million in investment for mitigation and adaptations measures per region.

These efforts culminated in the second Climate Leaders Summit in December of 2008, hosted by the Climate Group and chaired by Premier Rann of South Australia. The Climate Leaders Summit brought together sub-national efforts to focus on very specific sub-national partnerships to accelerate the low-carbon economy in several areas. These include energy efficiency such as green buildings, renewable energy, including solar and geothermal systems, low-carbon technology deployment in areas such as in electric vehicles and light-emitting diode (LED) lighting, sustainable land use such as agricultural best practice incentives, adaptation, and fiscal and regulatory measures to implement them. These partnerships are forming the basis for strategic exchanges throughout 2009 and beyond.

Conclusions: Linking the international and the national with the sub-national

In drafting a post-2012 climate change agreement, governments at the national level would do well to look towards the considerable work that is being undertaken at the state, provincial and regional levels. While progress at the international level may be slow, sub-national governments from around the world have been paving the way for action on climate change. States and regions are fast outpacing national governments in delivering effective and innovative climate policy. Pioneering climate change policies, sub-national governments are showing that a prosperous low-carbon economy is possible.

It is imperative that we negotiate an ambitious and effective international global climate change action plan and act now. Sub-national action is, and will remain, a necessary complement to the strategies and actions that already exist at the national level. It is important that heads of state, other world leaders and the UN itself recognize and support the role of sub-national governments in mitigation and adaptation to climate change. National governments will look to the efforts that states, provinces and regions have made in developing collaborative mechanisms in response to climate change as a model for national and international action.

President Obama is learning, no doubt, that a new day is dawning in the US approach to a changing climate. He is pointing to the economic opportunities that will arise from fast and bold climate change action, including job creation, generating growth and enhancing competitiveness. State-level action over the past few years shows what states are capable of in the absence of a constructive relationship with the federal administration regarding climate change. Now that the US has a federal administration that seems to be taking climate change seriously, there can be more optimism to what the next few years will bring.

21
A Green New Deal: Poverty Reduction and Economic Stability in a Carbon-Constrained World

Andrew Simms[1]

Introduction

Several years ago the International Red Cross sent me, on behalf of the *World Disasters Report*, to assess the early impacts of climate change upon vulnerable populations. What I saw in Tuvalu, in the South Pacific, and learned from other small island states about being resilient in the face of an unpredictable and extreme climate may hold lessons now for how many millions more can withstand the upheaval of global warming on our small island planet.

Tuvalu is living a uniquely modern paradox. It won the lottery of the internet age, being awarded the domain name '.tv'. Allegedly, it has a bigger delegation in Los Angeles to sell rights than it has at the United Nations to protect its political interests. Lying just a few metres above sea level, however, means that Tuvalu is in acute danger of losing its real home, just as it benefits from its new virtual one.

We can learn a lot from the mere fact that island communities such as Tuvalu have survived for so long on remote shards of land, exposed to the full force and vagaries of nature. To do so, they first had to respect their obvious environmental limits. They then evolved resilient local economies that helped them to cope with extreme and unpredictable weather. These were, of necessity, based on reciprocity, sharing and cooperation, and not unlimited growth fed by individualistic beggar-thy-neighbour competition.

Today, as collectively we face and exceed the limits of the Earth's bio-capacity, we are challenged at the global level to learn in a few short years lessons that such small communities often took millennia to arrive at. Our task is enormously complicated by the intricate interdependence of the modern global economy, the unbalanced distribution of power and benefits within it, and a pace of international decision-making that, until the ice started to melt so rapidly, I would have described as glacially slow. Fortunately, there is much that we already do know to guide our actions, drawing on decades of experience in dozens of countries and through thousands of community-based organizations around the world.

For example, in a series of reports looking in detail at different global regions, the Working Group on Climate Change and Development, a coalition of leading non-governmental organizations (NGOs) based in the UK that the New Economics Foundation (nef) helped to form, spelled out how climate change, if unchecked, stands not only to block further progress on the Millennium Development Goals (MDGs), but to reverse hard-won gains over many years. Our conclusion was that irreversible global warming, which appears perilously close, would mean not just greater hardship for millions, but the end of development as we have understood it for the last half a century (Simms et al, 2007).

One severe drought in Australia has already partly triggered worldwide food shortages and high and rising prices, creating shocks that ripple from the high street in the UK to the markets of Dhaka and Port au Prince. And the UK's official Hadley Centre for Climate Prediction and Research[2] recently concluded, based on a moderate scenario for change, that the percentage of the Earth's land surface prone to extreme drought having already trebled to 3 per cent in less than a decade will rise to fully one third by 2090, with droughts also longer in duration (Burke et al, 2006).

More worrying, still, the edge of the climate cliff is not clearly visible. Scientists such as the National Aeronautics and Space Administration's (NASA's) James Hansen[3] believe we may already be tipping over. This means not just stabilizing atmospheric greenhouse gases, but reducing them, with unimagined implications for the global economy. Misleadingly named 'positive environmental feedbacks' are volatile, hard to predict and may be terrifyingly sudden.

Due to the fact that the economy is a wholly owned subsidiary of the biosphere, we have no choice but to act, using precaution and the best information available. An individual may recover from financial bankruptcy; but if we allow our ecological debts to bankrupt a climate conducive to human civilization, geological history shows that it could take tens of thousands of years to be restored if, indeed, it ever is.

What, in this context, does precaution mean?

If we are lucky, 100 months from the beginning of August 2008, and based on a quite conservative estimate, we could reach a tipping point for the beginnings of runaway climate change (Simms, 2008).

The concentration of carbon dioxide in the atmosphere today, the most prevalent anthropogenic greenhouse gas, is the highest it has been for the past 650,000 years. In the space of just 250 years, as a result of the coal-fired

Industrial Revolution, and changes to land use, such as the growth of cities and the felling of forests, we have released, cumulatively, more than 1800 billion tonnes of carbon dioxide into the atmosphere. Currently, approximately 1000 tonnes of carbon dioxide is released into the Earth's atmosphere every second due to human activity. This is a figure that compares perversely to the fact that, during the first six months of 2008, the three fossil fuel companies BP, Centrica and Shell made the equivalent of UK£1000 profit every second (Lucas, 2008).

Greenhouse gases trap incoming solar radiation, warming the atmosphere. When these gases accumulate beyond a certain level – often termed a 'tipping point' – global warming will accelerate, potentially beyond control. A reasonable interpretation of precaution, in this case, would be to stay on the right side of that tipping point.

In climate change, a number of feedback loops amplify warming through physical processes that are either triggered by the initial warming itself, or the increase in greenhouse gases. One example is the melting of ice sheets. The loss of ice cover reduces the ability of the Earth's surface to reflect heat and, by revealing darker surfaces, increases the amount of heat absorbed. Other dynamics include the decreasing ability of oceans to absorb carbon dioxide due to higher wind strengths linked to climate change. This has already been observed in the Southern Ocean and North Atlantic, increasing the amount of carbon dioxide in the atmosphere and adding to climate change.

Because of such self-reinforcing positive feedbacks (which, because of the accidental humour of science, we must remind ourselves are, in fact, very negative), once a critical greenhouse concentration threshold is passed, global warming will continue even if we stop releasing additional greenhouse gases into the atmosphere. If that happens, the Earth's climate will shift into another more volatile state, with different ocean circulation, wind and rainfall patterns, the implications of which, according to a growing litany of research, are potentially catastrophic for life on Earth. Such a change in the state of the climate system is often referred to as irreversible climate change.

So, how exactly do we arrive at the ticking clock that takes us in 100 months from the summer of 2008 to the end of 2016? It is possible to estimate the length of time it will take to reach a tipping point. To do so you combine current greenhouse gas concentrations with the best estimates for the rates at which emissions are growing, the maximum concentration of greenhouse gases allowable to forestall potentially irreversible changes to the climate system, and the effect of those environmental feedbacks.

To make a reasonable, if cautious, estimate, it is possible to follow the latest data and trends for carbon dioxide, making allowances for all human interferences that influence temperatures, both those with warming and cooling effects. This estimate also follows the judgements of the mainstream climate science community, represented by the Intergovernmental Panel on Climate Change (IPCC), on what it will take to retain a good chance of not crossing the critical threshold of the Earth's average surface temperature rising by 2°C above pre-industrial levels. Even this steady approach, however, is perhaps too optimistic. Why? Because a rise of 2°C may mask big problems that begin at a lower level of warming. For example, collapse of the Greenland ice sheet is more than likely to be triggered by a local warming of 2.7°C, which could correspond to a global mean temperature increase of 2°C or less, and lead, over time, to a sea-level rise of up to 7m (IPCC, 2007).

Given all of the above, by the end of 2016 we stand to reach a concentration of greenhouse gases at which it is no longer 'likely' that we will stay below the 2°C temperature rise threshold. 'Likely' in this context refers to the definition of risk used by the IPCC. However, even just before that point, there is still a one third chance of crossing the line. Action, then, is more than urgent (IPCC, 2007).

We already know that people living in poverty are hit first and worst by global warming. This and the challenge of reducing poverty in a carbon-constrained world calls for a new development model that is both *climate proof* and *climate friendly*. From now on, all decisions will need to be scrutinized for whether they will increase or decrease vulnerability to climate change. We must look through the lenses of building resilience at the community level and reducing risk. And, it is the communities at risk who must shape our plans.

Parallel to the approach of the IPCC, the 2008 report of the International Assessment of Agricultural Knowledge, Science and Technology (IAASTD, 2008) showed that a massive shift of support to small-scale farmers using a diverse range of agro-ecological methods would be one of the most efficient ways to build resilience, inoculate against food crises and ensure against increasingly hostile weather patterns. Community-based coping strategies such as the use of seed banks, water management, vulnerability mapping, storm and flood protection that works with the local environment, and the conservation of forests and other ecosystems all represent effective ways for threatened communities to adapt.

If replicated and scaled up, small-scale renewable energy projects promoted by governments and community groups can help both to tackle poverty and reduce climate change. But this needs political commitment, significant new funds from governments and a major shift in priorities for energy lending by the World Bank and other development bodies. There is no either/or approach possible; the world must meet both its commitments to achieve the MDGs and tackle climate change. The two are inextricably linked.

Here we crash headlong into another equally large problem. It is clear that conventional economic growth will happen in poor countries as a consequence of effective poverty reduction. But at a global level, the policies designed to pursue growth have become a mask for making the rich richer, while leaving the poor with few benefits and abandoned to deal with growth's environmental consequences. During the 1980s, often referred to as the lost decade of development, for every US$100 worth of global economic growth, around US$2.20 (Simms and Woodward, 2006) found its way to people living below the absolute poverty line. A decade later that had shrunk to just US$0.60, and the actual mean income of those living under US$1 per day in Africa also fell (Chen and Ravallion, 2004).

There has been, in effect, a sort of 'flood up' of wealth from poor to rich, rather than a 'trickle down'. It means, perversely, that for the poor to get slightly less poor, the rich have to get very much richer, implying patterns of consumption that, in a world facing climate change, cannot be sustained. It now takes around US$166 worth of global growth (made up of all those energy-hungry giant flat screen TVs and sports utility vehicles) to generate a single dollar of poverty reduction for people in absolute poverty, compared with just US$45 in the 1980s. Earnings of between US$3 and US$4 per day are the approximate level at which the strong link between income and life expectancy breaks down. So, let us ask what would happen if we agreed US$3 per day as the minimum level of income to escape absolute poverty (Simms and Woodward, 2006).

Using the ecological footprint measure,[4] if the whole world wished to consume at the level of the US – a consumption pattern that has been fuelled, incidentally, by the credit binge that led to the current economic crisis – we would need, conservatively, over five planets like Earth to support us. But under the current pattern of unequally distributed benefits from growth, to lift everyone in the world onto a modest US$3 per day would require the resources of around 15 planets like ours. Where, you might ask, will the other 14 come from?

To tackle poverty in a carbon-constrained world we need a new development model, based on better measures of progress and a shift from relying on unequal global growth to serious redistribution. If we think of the planet as a cake, we can slice it differently; but we surely cannot bake a new one. Climate change is not the only reason that we have to learn to live with far fewer fossil fuels. Development must also contend with the volatility of oil prices, which reached record highs in 2008 and will probably return there as the world moves out of recession and the imminent global peak and long decline of oil production.

What, if any, guides do we have to survive these multiple shocks?

One country, much maligned, provides a glimpse of a near future that many more may face. Almost like a laboratory example, positioned on the flight path of the annual hurricane season, since 1990 Cuba has lived through the economic and environmental shocks that climate change and peak oil hold in store for the rest of the world.

The sudden loss of cheap Soviet oil and its economic isolation were so extreme at the end of the Cold War, and its reaction to the shock was so contrary to orthodox approaches and relatively successful, that it was dubbed in Washington the 'anti-model'. Then oil imports dropped by over half and the use of chemical pesticides and fertilizers dropped by 80 per cent. The availability of basic food staples such as wheat and other grains fell by half and, overall, the average Cuban's calorie intake fell by over one third in around five years (Pfeiffer, 2006).

Because of serious and long-term investment in science, engineering, health, education and land redistribution, there was reduced inequality. Research into low-input ecological farming techniques meant the country had a strong social fabric and the capacity to act.

At the heart of the transition after 1990 was the success of small farms and urban farms and gardens. The immediate crisis was averted by food programmes that targeted the most vulnerable people – the old, young, pregnant women and young mothers – and a rationing programme that guaranteed a minimum amount of food to everyone. Soon, half the food consumed in the capital, Havana, was grown in the city's own gardens. The threat of serious food shortages was overcome within five years (Novo and Murphy, 2001). Overall, urban gardens provide 60 per cent of the vegetables eaten in Cuba.

Time magazine recently called for a 'War on climate change' (*Time*, 2007), and, interestingly, Cuba's experience echoed what America achieved in a more distant time of hardship during World War II. Eleanor Roosevelt then led the 'victory gardening movement' to produce between 30 to 40 per cent of vegetables for domestic consumption, and public education campaigns warned that wasting fuel was like fighting for the enemy.

Cuba demonstrated that it is possible to feed a population under extreme economic stress with very few fossil fuels; but there were other surprises too. As calorie intake fell by more than one third, of necessity the proportion of physically active adults more than doubled and obesity halved. From 1997 to 2002, deaths attributed to diabetes halved, coronary heart disease fell by 35 per cent, and strokes and other causes by around one fifth (Franco et al, 2007). The approach was dubbed the 'anti-model' because it was both highly managed and led by communities, it focused on meeting domestic needs rather than exports, was largely organic and was built on the success of small farms (Pfeiffer, 2006).

The same country's approach to disaster preparedness and management is also instructive. Compared to the deaths and destruction in New Orleans following Hurricane Katrina, when Hurricane Michelle hit Cuba in 2001, only five lives were lost and recovery was quick. It was due to proper planning and a collective approach managed by government, but owned at the local level. Disasters expert Ben Wisner commented on the evacuation of 700,000 of Cuba's 11 million people:

> *This is quite a feat given Cuba's dilapidated fleet of vehicles, fuel shortage and poor road system. At least one analyst suggests that the Cuban experiment 'may hold many of the keys to the future survival of civilization'.* (Wisner, 2001)

Currently, according to our calculations, in a given calendar year the world as a whole goes into ecological debt around 7 October – by which time we have consumed more and produced more waste than ecosystems can deal with. The results are seen in climate change, oceans emptied of fish and desertification. Forty years ago Senator Robert Kennedy said that:

> *Gross National Product does not allow for the health of our families, the quality of their education, or the joy of their play. It is indifferent to the decency of our factories and the safety of our streets alike. It does not include the beauty of our poetry, or the strength of our marriages, the intelligence of our public debate or the integrity of our public officials... The Gross National Product measures neither our wit nor our courage, neither our wisdom nor our learning, neither our compassion nor our devotion to our country. It measures everything, in short, except that which makes life worthwhile.* (Kennedy, 1968)

But it is possible to assess if we are achieving human development while living within our environmental means.

Nef's own Happy Planet Index compares the relative success of nations at delivering long life expectancy and high levels of well-being, compared to their size of ecological footprint. The results reveal many middle-income countries performing well, with good life expectancy and well-being and relatively low footprints. Strikingly, some of the best performers are small island states. Somehow, they have worked together to produce more convivial communities while respecting environmental limits.

The UN faces huge challenges. Not least is how to recognize and protect the large and growing number of people we can expect to be displaced in a warming world. The climate refugee crisis will dwarf that of political refugees. What will happen to the nationhood and economic areas of countries that could disappear entirely, such as Tuvalu? How can we change our locked-in thinking about economic development and reorganize around the principles of resilience, social justice, sufficiency, ecological efficiency and the capacity to adapt?

We might begin by asking, as acid tests:

- Will what we do make people more or less vulnerable?
- Will it move us toward truly sustainable one-planet living?
- Will it move us fast enough to prevent irreversible, catastrophic climate change?

When the people of Tuvalu first encountered Europeans in the 19th century, they gave them the name Palangi. Victorian travellers translated the word to mean 'heaven bursters', a reference to their ship's guns. Now, some of our lifestyles truly threaten to burst the heavens. At the very least, to achieve poverty reduction in a world threatened by climate change, we know that rich countries must radically cut their own consumption to free up the environmental space in which others can pursue, as a first step, the Millennium Development Goals.

The good news is that we now know from the literature on human well-being that making the rich richer does nothing to increase their life satisfaction. On the contrary, numerous studies confirm that once your basic needs are met, you are just as likely to have high life satisfaction, whether your ecological footprint is large or small. My conclusion is that a new development model is needed as much, if not more, in countries such as the UK and the US as in the majority developing world. It is us who have to demonstrate that good lives do not have to cost the Earth.

Given the speed with which we must now act, perhaps there is a lesson that we can learn from the past. At the time of writing, the nations of the UK and the US are gripped by a credit crisis (although more properly called a 'debt' crisis), threatening recession or even full-blown depression, and the impact of high oil and rising food prices, as well as the massive added challenge of having to avert climate change.

During the 1930s President Roosevelt launched his New Deal[5] to tackle the excesses of a reckless domestic financial sector and to lay the foundations of economic recovery, which also put the US in a position, a few years later, to enter and help win World War II.

Does a war comparison sound dramatic?

In April 2007, Margaret Beckett, then foreign secretary, gave a largely overlooked lecture called 'Climate change: The gathering storm':

> *It was a time when Churchill, perceiving the dangers that lay ahead, struggled to mobilize the political will and industrial energy of the British Empire to meet those dangers. He did so often in the face of strong opposition. Climate change is the gathering storm of our generation. And the implications – should we fail to act – could be no less dire: and perhaps even more so.* (Beckett, 2007)

So, Britain and the US could launch a Green New Deal (Simms et al, 2008) for 100 months of concerted action, taking inspiration from Roosevelt's famous 100-day programme to implement his New Deal in the face of the dust bowls and depression. One such plan we published in the UK by a group of specialists in finance, energy and environmental specialists.

Addressed at the triple crunch of the credit crisis, high oil prices and global warming, the basics of the plan are to rein in reckless financial institutions and employ a range of fiscal tools, new measures and reforms to the tax system, such as a windfall tax on oil companies. The resources raised are then to be invested in a massive environmental transformation programme that could insulate the economy from recession, create countless new jobs and allow those countries to play their part in meeting the climate challenge.

As a precursor to enabling and building more sustainable systems for transport, energy, food and overhauling the nation's building stock, the governments need to brace themselves to tackle the big financial institutions. Currently they are giving us the worst of all worlds. We have woken to find the foundations of our economy made up of unstable, exotic financial instruments. At the same time, and perversely, as awareness of climate change goes up, ever more money pours through the City, London's financial district, into the oil companies. These companies list their fossil fuel reserves as 'proven' or 'probable'. A new category of 'un-burnable' should be introduced to fundamentally change the balance of power in the City. Instead of using vast sums of public money to bail out banks because they are considered 'too big to fail', they should be reduced in size until they are small enough to fail without hurting anyone. It is only a climate system capable of supporting human civilization that is too big to fail.

In 2008 we saw considerable oil price volatility. There is considerable potential to tax oil companies at a higher rate. Many politicians argued for this so-called 'windfall tax'. Money raised – in this way and through other changes in taxation, new priorities for pension funds and innovatory types of bonds – would go towards a long overdue massive de-carbonization of our energy system. Decentralization, renewables, energy efficiency, conservation and demand management will all play a part.

Next would be a rolling programme to overhaul the nations' heat-leaking building stock. This will have the benefit of massively cutting emissions and at the same time tackling the sore of fuel poverty by creating better insulated and designed homes. A transition from 'one person, one car' on the roads to a variety of clean reliable forms of public transport should be visible by the middle of our 100 months. Similarly, weaning agriculture off fossil fuel dependency will be a phased process.

The end result will be real international leadership, removing the excuses of other nations not to act. But it will also leave people more secure in terms of food and energy supplies, and with more resilient economies capable of weathering whatever economic and environmental shocks the world has to throw at us. Each of these challenges will draw on things that we already know how to do, but have missed the political will for.

Impassable ecological obstacles lie on the path down which we chase the shadows of overconsumption to deliver our well-being, expecting the poor to be grateful for the crumbs that fall from our plates. The good news is that another way is not only possible, as the philosopher A. C. Grayling writes: 'it is better, richer and more enduring'.[6]

Notes

1 This chapter is expanded from a speech given by Andrew Simms, policy director and head of the Climate Change Programme at the New Economics Foundation, to the UN ECOSOC special session on climate change and the MDGs, New York, 2 May 2008.

2 Because we can't know the future for certain, our climate change scientists use computer-based climate models to project plausible scenarios, or projections, for coming centuries. It is important to be aware that projections from climate models are always subject to uncertainty because of limitations on our knowledge of how the climate system works and on the computing resources available. Different climate models can give different projections. See more on using computer models at www.metoffice.gov.uk/climatechange/science/projections/.

3 Hansen's testimony on climate change to congressional committees during the 1980s helped to raise broad awareness of global warming and his continuing advocacy for the science of climate change has been influential in shifting the policy debate in the US and globally. He has been a critic of both the Clinton and Bush administrations' position on climate change.
4 The ecological footprint is a measure of human demand on the Earth's ecosystems. It compares human demand with planet Earth's ecological capacity to regenerate. It represents the amount of biologically productive land and sea area needed to regenerate the resources that a human population consumes and to absorb and render harmless the corresponding waste. Using this assessment, it is possible to estimate how much of the Earth (or how many planet Earths) it would take to support humanity if everybody lived a given lifestyle. Humanity's total ecological footprint is estimated at 1.3 – in other words, humanity needs 1.3 Earths in order to sustain our collective lifestyle. This is just one of many varying figures on the global ecological footprint.
5 The New Deal was the name that US President Franklin D. Roosevelt gave to a sequence of programmes he initiated between 1933 and 1936 with the goal of giving work (relief) to the unemployed, reform of business and financial practices, and recovery of the economy during the Great Depression.
6 A. C. Grayling's main areas of interest in technical philosophy lie at the intersection of theory of knowledge, metaphysics and philosophical logic. He brings these subjects together in an attempt to define the relationship between mind and world, and in so doing he is, among other things, challenging philosophical scepticism.

References

Beckett, M. (2007) 'Climate change: The gathering storm', Speech, www.link.brightcove.com/services/player/bcpid769460422?bclid=769311289&bctid=769921003

Burke, E. J., S. J. Brown and N. Christidis (2006) *Modelling the Recent Evolution of Global Drought and Projections for the Twenty-First Century with the Hadley Centre Climate Model*, Hadley Centre for Climate Prediction and Research

Chen, S. and M. Ravallion (2004) *How Have the World's Poorest Fared Since the Early 1980s?*, Development Research Group, World Bank, Washington, DC

Franco, M., R. Ordunez, B. Caballero, J. A. T. Granados, J. L. Bernal, E. Guallar and R. S. Cooper (2007) 'Impact of energy intake, physical activity and population-wide weight loss on cardiovascular disease and diabetes mortality in Cuba, 1980–2005', *American Journal of Epidemiology*, vol 166, no 12, pp1374–1380

IAASTD (International Assessment of Agricultural Knowledge, Science and Technology) (2008) *Summary for Decision Makers*, Island Press, Washington, DC, www.agassessment.org/index.cfm?Page=IAASTD%20Reports&ItemID=2713

IPCC (Intergovernmental Panel on Climate Change) (2007) *Fourth Assessment Report*, IPCC, Geneva, www.ipcc.ch/ipccreports/assessments-reports.htm

Kennedy, R. (1968) Speech at University of Kansas, www.coolmelbourne.org/the_important_stuff/renew_your_outlook/RF_Kennedy_on_GDP.html

Lucas, C. (2008) *Green New Deal to Overcome Credit Crunch*, Green Party, London, www.carolinelucas.com/?q=node/55

Novo, G. M. and C. Murphy (2001) 'Urban agriculture in the city of Havana: A popular response to a crisis', in *Growing Cities Growing Food: Urban Agriculture on the Policy Agenda: A Reader on Urban Agriculture*, Resource Centres on Urban Agriculture and Food Security, www.ruaf.org/node/82

Pfeiffer, D. A. (2006) *Eating Fossil Fuels – Oil, Food and the Coming Crisis in Agriculture*, New Society Publishers, Canada

Simms, A. (2008) 'The final countdown', *The Guardian*, London, August, www.guardian.co.uk/environment/2008/aug/01/climatechange.carbonemissions

Simms, A. and D. Woodward (2006) *Growth Isn't Working: The Uneven Distribution of Benefits and Costs from Economic Growth*, New Economics Foundation, London, www.neweconomics.org/gen/z_sys_publicationdetail.aspx?pid=219

Simms, A., J. Magrath and H. Reid (2007) *Up in Smoke? Threats from, and Responses to, the Impact of Global Warming on Human Development*, IIED, London, www.iied.org/pubs/display.php?o=9512IIED

Simms, A., A. Pettifor, C. Lucas, C. Secrett, C. Hines, J. Leggett, L. Elliot, R. Murphy and T. Juniper (2008) *Green New Deal*, New Economics Foundation, London, www.neweconomics.org/gen/z_sys_publicationdetail.aspx?pid=258

Time (2007) 'How to win the war on global warming', *Time*, Washington, DC, www.time.com/time/specials/2007/article/0,28804,1730759_1731383_1731363,00.html

Wisner, B. (2001) 'Socialism and storms', *The Guardian*, London, 14 November, www.guardian.co.uk/world/2001/nov/14/cuba.socialsciences

22
Energy, Climate Change and Human Security: The Trade Union Challenge

Winston Gereluk and Lucien Royer

Introduction

Trade unions bring a worker's perspective to the discussion of human security and climate change. Not only have workers and their communities experienced first hand the death, suffering and physical devastation caused by weather events over the last decade; they have also borne the long-term loss and disruption to jobs and livelihoods.

Whether it was the 2004 tsunami that devastated countries over two continents, killing thousands of people and displacing over 1 million more, the 2003 hurricanes that destroyed the south-eastern coast of the US, or the mudslides in the Philippines and El Salvador, millions of people have lost their livelihoods because of severe weather events and the number will only rise if scientific predictions come true. To workers and their families, these losses are, in many ways, a greater threat than military aggression or terrorism on which dominant concepts of security have been based. For them, energy, human security and climate change are closely interlinked.

However, workers cannot be simply regarded as victims. Because they are at the centre of the industrial activity that is putting so much stress on our planet's ecosystems, they are in a good position to contribute to climate change measures and work for clean energy solutions.

By linking climate change to issues of human security, the April 2007 debate in the United Nations Security Council marked a watershed in international responses to climate change. It also entrenched an earlier UN definition of threat to international security as 'any event or process that leads to large-scale death or lessening of life chances and undermines states as the basic unit of the international system', (Wisner et al, 2006), specifically highlighting economic and social threats. This corresponds to a concept of human security that has been promoted by trade unions and underlined a UN disaster reduction report finding that the number of people affected by natural disasters between 1990 and 1999 was six times more than those affected by armed conflicts (UNISDR, 2003).

Social effects of climate change: Livelihood contraction and job loss

Trade unions have focused on the social dimension ever since they joined international deliberations on sustainable development in the 1980s, led by the International Confederation of Free Trade Unions (ICFTU) – now the International Trade Union Confederation (ITUC) – the Trade Union Advisory Committee (TUAC) to the Organisation for Economic Co-operation and Development (OECD) and certain Global Union federations. It is from this perspective that they address human security issues related to the growing crises in energy, climate change and human security, and why they support the multidisciplinary concept underlying the UN Secretary General's 2005 report *Larger Freedom* (Annan, 2005), according to which people can only be secure if they are protected from possible threats and able to act on their social, economic and environmental rights (UNDP, 1994).

Jon Barnet and Neil Adger have added a useful perspective in a paper for a 2005 workshop on Human Security and Climate Change in Oslo, Norway. Climate change threatens human security, they said, 'principally through its potentially negative effects on people's livelihoods' (Barnet and Adger, 2005). Such 'livelihood contraction', often related to declining access to natural capital caused by, for example, deforestation, land degradation, drought, agricultural expansion or population displacement by dams, etc. usually implies a loss of gainful employment (Barnet and Adger, 2005, p6).

From the earliest days of the United Nations Framework Convention on Climate Change (UNFCCC) and the Kyoto Protocol, trade unions have focused on the employment and the insecurity created by current patterns of energy use, as well as by proposed climate change measures. While most studies suggest some positive net growth in world employment in the long term, when job losses are measured against job gains, sectoral and regional analyses of climate change effects show that large numbers of workers will lose jobs and livelihoods in the short term, with little access to new jobs. Unfortunately, even though this source of insecurity is likely to be significant, it has been largely absent from international research and debate. Even Stern did not specifically refer to employment impacts in his list of 'highlights of possible climate impacts' (Stern et al, 2006).

After a decade of asking for cooperation from such intergovernmental organizations as the OECD, the International Labour Organization (ILO) and the European Union (EU) to study these linkages,[1] a study was undertaken by the European Trade Union Confederation (ETUC), Syndex, the German Wuppertal Institute and the Spanish institute ISTAS (Instituto Sindical de

Trabajo, Ambiente y Salud), and financed by the European Commission and Belgium, Finland, France, Italy, Spain and the UK. The report *European Governments and Unions Assess Employment and Climate Change Synergies* (ETUC et al, 2005) examined the potential impact upon employment of the carbon dioxide reduction targets of 30 to 50 per cent by 2030 proposed by the European Union on four key industrial sectors: energy production (electricity production and oil), energy-intensive industries (steel and cement), transport and building/construction, and concluded that there could be an overall net gain in employment in the order of 1.5 per cent in sectors identified with transfers to environmentally friendly forms of energy production and use. It warned, however, that jobs will only be created where companies take advantage of opportunities created by climate policies and lost where companies cannot adapt (ETUC et al, 2005).

At the latest Conference of the Parties (COP) in Nairobi in 2006, the ITUC renewed its call for more research to establish climate change linkages to employment, as well as to sustainable development, bringing into the UNFCCC the discussion of social issues related to adaptation, mitigation and the various Kyoto mechanisms (Global Unions, 2006a). The ILO Labour Conference that followed in June 2007 concluded that 'it is essential that the employment and income dimensions are taken into account in the post-Kyoto agreements and in the implementation of programmes. This requires a much better understanding of the links and that employers and workers participate in making decisions and in building transitions to the low carbon economies of the future'(ILO, 2007).

Far-reaching as they may be, the reality is that climate change would only add to the insecurity of precarious employment under which millions of the world's poor already work. The Millennium Development Goal (MDG) of decent, sustainable employment is today a fiction for one third of the world's workers who are unemployed or relegated to short-term, part-time, unsafe jobs with low wages and long hours, or work in the informal economy with no rights or social protection. In recent decades, a dramatic increase in this type of employment in both developing and industrialized countries has accompanied a shift from manufacturing to service industries, a spread of new information and communications technologies, and labour processes that emphasize flexible employment relations (Fudge and Owens, 2009, p401). Women bear the brunt of this insecurity, as do millions of children who are condemned to child labour (ILO, 1995).

Poverty is at the root of this insecurity, of course; but in some countries and industries, it is actively fostered by economic and social policies of the government and the inattention of governmental regulatory agencies (Warshaw, 2007). The International Metalworkers Federation (IMF) is one of a several trade union federations to note an increase in the number and types of precarious employment situations – specifically, direct hire of temporary labour contracts, hiring by employment agencies or brokers, contracting out, personal labour contracts as bogus 'self-employed' workers, disguised employment training contracts, and home working (IMF, 2007).

Human security, poverty and vulnerability

Climate change does not by itself undermine human security or increase risk of conflict; these usually depend upon a combination of 'vulnerability' factors. The United Nations Development Programme (UNDP) has produced measures for these linkages in its Human Development Index (HDI), which correlates with its Disaster Risk Index (DRI), especially for less developed countries (UNDP, 2004). Not surprisingly, these show that climate change is likely to create or exacerbate human insecurity where livelihoods are already vulnerable, poverty levels are high, states are weak, extensive human migration exists, or where gender issues play a significant role (Delaney and Shrader, 2000).

These factors explain why climate change and other environmental effects hit hardest at the world's poorest inhabitants, including the 50 million 'environmental refugees' whom the United Nations University's Institute for Environment Security estimates will be displaced in the next five years. In its *Fourth Assessment Report*, the IPCC relates human security to basic needs addressed in the Millennium Development Goals for food, water, health and a place to live. Poor communities, it said, can be especially vulnerable as they tend to have more limited adaptive capacities, and are more dependent upon climate-sensitive resources such as local water and food supplies (IPCC, 2007).

Human security, therefore, raises issues of equity, political economic power, human rights and sustainable development. People do not contribute equally to the causes of climate change; nor are they equally capable of shaping responses or adapting to its effects. Questions such as 'whose security is threatened?' or 'how are responses constrained or facilitated by existing institutions and power structures?' are exemplified in large parts of Africa, 'one of the most vulnerable continents to climate variability and change because of multiple stresses and low adaptive capacity' (IPCC, 2007, p10). In the debate over climate change, we must constantly be reminded that much of the world does not require weather events to experience human insecurity. For the nearly one half of the global population – 3 billion people – who live on less than US$2 a day, and the one fifth (1.5 billion) who live in extreme poverty, existence is already precarious and human security a dream.

As the *Human Development Report* points out, however, 'Human poverty is more than income poverty – it is the denial of choices and opportunities for living a tolerable life' (Baker Institute and Energy Forum, 2006). Lack of access to energy services for the poor poses a serious human security challenge, as it renders basic social goods such as healthcare and education even more costly and inaccessible. Substantial social and economic development simply cannot occur, and MDGs cannot be met without major improvement in the quality and quantity of energy services in developing countries (UNDP, 2006).

Even though electricity supply has been extended to 1.3 billion people in developing countries in the last 25 years, 1.4 billion people will '*still*' lack access to electricity, while the number reliant on biomass will increase from 2.5 billion in 2006 to 2.7 billion by 2030 (Baker Institute and Energy Forum, 2006).

According to the World Health Organization (WHO), exposure to indoor air pollution as a result of biomass is responsible for the nearly 2 million excess deaths, primarily women and children, from cancer, respiratory infections and lung diseases and for 4 per cent of the global burden of disease. In relative terms, biomass pollution kills more people than malaria and tuberculosis each year (Bruce et al, 2000).

Thus, while the end of 'cheap oil' may stimulate alternative energy and conservation, it is likely to tax local and national capacity that is already over-extended, particularly for the poor, according to the International Federation of Chemical, Energy, Mine and General Workers' Unions (ICEM). 'Security of supply and sustainable development', it says, 'remain at the heart of the political debate around energy' (ICEM, 2006). In fact, spiralling oil prices have become a growing source of poverty for many of the world's poorest countries that depend on energy imports, including top recipients of debt relief under the Heavily Indebted Poor Countries Initiative, which are forced to spend resources on oil rather than meeting such needs as education, improving access to health services needed to counteract HIV/AIDS, improving water quality or enhancing their infrastructure (Roche and Perez, 2006).

To this injustice can be added the fact that the global dependence on fossil fuels is a major cause of the global climate change that is already triggering weather events that can wipe out local economies and kill thousands of people, particularly in poverty-ridden communities that often do not have immediate access to relief services or adequate resources to recover from disasters, not to mention the threats it poses to food production and human health (McMichael et al, 2003). As well, the poor have always been most vulnerable to the boom-and-bust cycles of energy resource-dependent economies. Instead of benefiting from their energy resources, they bear the brunt of oil spills and inadequate public services, as in the Niger Delta, where communities near oil facilities typically lack electricity and running water, or in Aceh, Indonesia, where natural gas exports were interrupted by protests and risk of violent conflict among indigenous people – another side of unsustainable energy policies and practice (Baker Institute and Energy Forum, 2006).

Trade union responses to climate change and human insecurity

The United Nations Security Council debate occurred at a time when many of the world's trade unions recognized that climate change and energy security must form part of their historical mission to ensure that development yields personal and community benefits to the working class. In response to calls for action issued by the ITUC, TUAC and several Global Union federations, a Trade Union Working Group on Climate Change has been formed, which now enjoys official status at negotiations under the UNFCCC.

In January 2006, trade unions partnered with the United Nations Environment Programme (UNEP) to convene the first Global Trade Union Assembly on Labour and the Environment in Nairobi, Kenya, from which over 200 trade union leaders and activists from all parts of the world issued a statement calling for action linking poverty reduction, environmental protection and decent work (Global Unions, 2006b). This was followed later that year by regional conferences in Brazil and South Africa organized by Sustainlabour, an international foundation that promotes sustainable development in the trade union movement.

On 7 to 8 May 2007 a North American Labor Assembly on Climate Crisis at Cornell University's Global Labor Institute in New York City brought hundreds of trade unionists from around the world, together with representatives from the energy policy community and environmental organizations, to address energy alternatives, job creation, policy options and coalition-building. It adopted a statement urging 'determined action to address the climate crisis, a crisis that threatens life on our planet, as we know it'. Inaction, it said, 'is a far greater threat to workers and communities than is taking decisive action now and in the years ahead' (North American Labor Assembly on Climate Crisis, 2007).

The emerging trade union consensus firmly rejects the 'jobs/environment dichotomy'. As the United Steel Workers Association (USWA) states in *Securing Our Children's World*: 'In the last five years, we can clearly see that the 3 million manufacturing jobs that vanished from the US economy had almost nothing to do with domestic environmental regulation, and everything to do with US corporate trade policy' (USWA, 2002, p29). As well, it rejects solutions based on the model of privatization and government cutbacks that prevailed during the last decade. An officer of Public Services International (PSI) made the point to the 2007 Session of the Commission on Sustainable Development:

> *Ever since the Enron nightmare, public decision-makers have been scrambling to fix the problems of energy privatization, deregulation and liberalization. The failure of such policy approaches has been well documented, and [these approaches] have been particularly egregious when they are intended to either reduce government debt or reduce corruption. The private sector has an important role to play; but only where they do more than provide a temporary reduction of government borrowing and able to deliver equitable and safe energy services to all. Too often, they have failed our citizens and our planet, particularly, especially in developing countries, where we see major global corporations abandon their commitments when profits are too low or risks too high.* (Boys, 2005)

A just transition to an economy based on sustainable energy

Decent employment is the only reliable route to human security for millions of the world's poor. Unfortunately, most mitigation and adaptation measures, including a change to a sustainable energy economy, will have human security effects, decreasing vulnerability for some but increasing it for others who lack access to financial resources and technology, education and the key institutions – particularly political representation and access to power (O'Brien et al, 2007). As the 2006 Nairobi conference noted, however, 'Creating decent and secure jobs is possible only if environmental sustainability is attained; hence the need to embrace the poverty reduction and sustainable development goals contained in the Millennium Declaration and Johannesburg Plan of Implementation through the promotion of decent employment and environmental responsibility' (Global Unions, 2006b).

Such issues relate to the 'biofuel solution' to energy demand as large-scale investments are already taking land out of food production and diverting food grains, raising food prices and eroding biodiversity (McNeely, 2006).

They may also apply to some Clean Development Mechanisms (CDMs) under Kyoto that increase dislocation and exclude marginal sections of host communities, particularly large-scale water management and forestry projects (World Commission on Dams, 2000).

Since the earliest conferences and meetings of subsidiary bodies of the UNFCCC, trade unions have consistently lobbied for 'just employment transition policies' so that workers would not have to suffer economic hardship or insecurity as a result of climate change or other environmental measures.

The ETUC renewed the call for adequate funding, negotiated with social partners to ensure that workers are both able to adapt and offered security, to include, as a minimum, adequate job forecast and skills management, a reconversion programme for workers from affected industries, and an income-support programme for periods of unemployment (ETUC, 2006, p14). Governments, it said, must take the lead, beginning with ratification of such ILO Conventions as C-122, the Employment Policy Convention or the 1974 ILO Resolution on the Social and Economic Consequences of Preventive Action, which calls for special arrangements for socio-economic consequences affecting workers.

The call for a 'just transition' has been taken up wherever trade unions joined the debate on climate change. Not only would such policies ensure an element of justice, they would build confidence and support for national efforts to address climate change (Global Unions, 2006b). As noted in their Nairobi Action Plan for Africa, action to engage workers and their communities in the development of climate change adaptation strategies is urgently required on that continent (Global Unions, 2006b).

The role of good industrial relations

Implementation of climate change measures invariably involves the world's workplaces, the hub of production and much of the world's consumption. Effective change, however, will require the involvement of workers and their trade unions, working with employers to change the way in which they burn energy, consume resources and generate waste, as well as to affect personal patterns (e.g. the way in which they travel to and from work). Workers can only be confident that their interests are being taken into account if they play a role in planning and implementation, with full accountability and transparency – hence the need for industrial relations models based on participation and dialogue, as contemplated in Chapter 29 of Agenda 21, which defined a role in sustainable development for workers and trade unions.[2]

The most effective trade union mechanism for this purpose is the workplace committee for environment or occupational health, whose primary mandate is to educate members, as well as undertake such activities as research and monitoring of company performance. Article 6 of the New Delhi Programme of Work asks parties to advance and report on activities for public awareness through education and training as a route to public participation and access to information, and, in many countries, trade unions are in the best position to contribute, as they are the largest single provider of informal adult education.

The USWA also points out that:

> *Workers are in a key position in the fight for environmental quality. Violations of pollution regulations can be difficult for the public to spot. Nor is it possible for the government to monitor continuously every potential polluter. It is much harder to hide illegal or other inappropriate behaviour from plant workers. And through collective bargaining and the power of the union, organized workers have an especially effective tool for ensuring cleanup or sustainable forms of production.* (USWA, 2002, p34)

Dialogue is also the key ingredient in over 2 million collective agreements negotiated by trade unions worldwide. These are supplemented by European works councils and the framework agreements that provide a platform of standards that a multinational company must apply wherever it operates in the world, even where national law is lacking. Recently in Germany, the works council has provided the right of workers to request and obtain information pertaining to the environmental performance of their enterprise. One of the earliest, between the ICEM and Statoil in 1998, was endorsed by the company vice president who acknowledged unions as 'globe-spanning knowledge-based organizations. They give us early warning of problems we should be aware of, and allow us to take early action to mitigate risks.' Companies that take a cooperative rather than adversarial stance towards civil society and labour, he said, 'have history on their side' (ICEM, 2006).

Dialogue on workplace action is being actively promoted at the level of the plant, the sector and nationally. The Trade Union Advisory Committee to the OECD is committed to work with the business arm of the OECD on climate change and energy concerns, as well as with the OECD Environmental Policy Committee (EPOC). The ETUC study recommends an extension of this type of dialogue to include a tripartite European dialogue on implementation of adaptation and mitigation based on the model of Kyoto round tables in Spain, as well as a European Observatory on the economic and social upheavals linked with climate change to be tasked with supporting the development of industrial relations in this new area (ETUC, 2006).

Strategic global alliances and cooperation to address climate change

Trade unions attribute many issues concerning energy, climate change and human security to deepening globalization, a condition that calls for a collective global response. National efforts towards sustainable development can only succeed if, for instance, trade in harmful products is restricted or if Multilateral Environmental Agreements prevent overexploitation of natural resources for export. In his report, Stern pointed to examples of successful collective action in the areas of international trade, health, development aid, terrorism and environmental protection as cause for hope (Stern et al, 2006, p453).

To this end, trade unions are creating strategic alliances with business and other Agenda 21 major groups, such as the Climate Action Network, with whom they are developing joint plans of work for meetings of COPs and subsidiary bodies of the UNFCCC. They also support such compliance systems as the Kyoto Protocol, and are providing input through national governments to Article 9 reviews for the post-Kyoto terms of reference (Nieto, 2006).

Trade unions have never subscribed to the popular notion that globalization renders nations incapable of serving the needs of their citizens. They see the state as playing a central role in providing opportunities for people, providing a stable environment so that livelihoods can be pursued with confidence, and providing measures to protect people whose livelihoods are threatened and ensure political freedoms, as well as providing such social opportunities as education and healthcare, etc. As one example, national governments can do much to promote 'green jobs' linked with 'decent work', as urged by ILO Secretary General Juan Somavia in his keynote address to the 2007 ILO Labour Conference. He urged government, worker and employer delegates to develop policy tools through a tripartite social dialogue framework for a global Green Jobs Initiative to support workers and enterprises through the transition to a much more environmentally sustainable process of development (Somavía, 2007).

To this end, trade unions will pursue national and sectoral tripartite approaches, such as the 2005 agreement between the Spanish government, the Confederación Sindical de Comisiones Obreras (CCOO) and Unión General de Trabajadores (UGT) and leading business organizations. Parties were brought together around a Dialogue Table to guide national efforts to meet obligations under the Kyoto Protocol, with specific reference to social and employment impacts. It was followed in June 2005 by an agreement between the government of Argentina, the Confederación General del Trabajo (CGT) and employers to engage in joint target-setting, monitoring, record-keeping and implementation of management systems. As well, the government of Belgium has geared its approach to 'flexible mechanisms' under Kyoto to social and employment policy. Project proposals must apply the principles of the OECD's *Guidelines for Multinationals*, ILO conventions and the Declaration on Fundamental Principles and Rights at Work, and trade unions are involved in monitoring to ensure essential services, including energy, to local populations and performance on environmental, social and economic impacts.[3]

The Trade Union Sustainable Development Unit has been set up to track and report on the performance of states along energy and climate indicators, a process now mainstreamed by a number of United Nations agencies.[4] As well, unions will press for vigorous national programmes through 'blue–green' alliances between unions, and environmental and community organizations, such as the new Apollo Project in the US, which intends to forge a national commitment that mobilizes resources to achieve a new energy infrastructure that is diversified, environmentally safe and more efficient, to generate good jobs and to help capture the green markets of the future (New Energy for America, 2006).

Notes

1 ETUC (European Trade Union Confederation) (2005) *European Governments and Unions Assess Employment and Climate Change Synergies*: a ten-month study on the impacts of climate change and carbon dioxide reduction policies upon employment and other activities in 25 EU countries conducted under the auspices of the European Agency Social Development Agency (SDA) and the European Trade Union Confederation.

2 Industrial relations refers to the rules governing relations between employers and employees (or their representatives), premised on the understanding that the needs and aspirations of both parties must be addressed and, further, that only trade unions can provide representation that is regarded by workers as authentic and reliable.

3 See Belgium's approach to buying carbon credits: www.klimaat.be/jicdmtender/indexB.htm

4 Information on what the trade unions are doing can be found at www.un.org/esa/sustdev/mgroups/tu_cprofiles.

References

Annan, K. (2005) *Larger Freedom: Towards Development, Security and Human Rights for All*, United Nations, New York, NY, www.un.org/largerfreedom/

Baker Institute and Energy Forum (2006) *Poverty, Energy and Society*, www.rice.edu/energy/research/poverty&energy/

Barnet, J. and N. Adger (2005) *Security and Climate Change: Towards an Improved Understanding*, Human Security and Climate Change Workshop, Oslo, Norway, p15

Boys, D. (2005) Speech to the 14th Session of the Commission on Sustainable Development, New York, NY, May

Bruce, N., R. Perez-Padilla and R. Albalak (2000) 'Indoor air pollution in developing countries: A major environmental and public health challenge', *Bulletin of the World Health Organization,* WHO, Geneva, pp1078–1092

Delaney, P. L. and E. Shrader (2000) *Gender and Post Disaster Reconstruction*, LCSPG/LAC Gender Team, World Bank, Washington, DC, p24, www.gdnonline.org/resources/reviewdraft.doc

ETUC (European Trade Union Confederation) (2006) *Climate Change and Environment*, www.tradeunionpress.eu/Web/EN/Activities/Environment/Studyclimatechange/rapport.pdf

ETUC (European Trade Union Confederation), SYNDEX, the German Wuppertal Institute and ISTAS (Instituto Sindical de Trabajo, Ambiente y Salud) (2005) *European Governments and Unions Assess Employment and Climate Change Synergies*

Fudge, J. and R. Owens (2006) *Precarious Work, Women, and the New Economy: The Challenge to Legal Norms*, Hart Publishing, Oxford, UK, p401

Global Unions (2006a) *Report of Trade Unions at the UN Framework Convention on Climate Change – UNFCCC COP12/MOP2*, 6–17 November 2006, Nairobi, Kenya, www.global-unions.org/pdf/ohsewpP_8Bb.EN.pdf

Global Unions (2006b) *Unions Call For Solidarity Over Climate Change Action for Africa*, Trade union statement to Nairobi Climate Change Conference, COP 12 – MOP 2, www.global-unions.org/pdf/ohsewpP_8Ae.EN.pdf

ICEM (International Federation of Chemical, Energy, Mine and General Workers' Unions) (2006) *World Conference for Energy Industries*, November 2006, www.icem.org/index.php?id=166&doc=2060&la=EN

ILO (International Labour Organization) (1995) *Report of the ILO Committee on Employment and Social Policy: Child Labour*, GB.264/ESP/1, 264th Session, ILO, Geneva

ILO (2007) *Impact of Climate Change on Employment, Incomes and Poverty*, International Research Workshop, International Institute for Labour Studies and Policy Integration, Geneva, Switzerland

IMF (International Monetary Fund) (2007) 'Global action against precarious work', *Metal World*, vol 18, no 1, IMF, Washington, DC

IPCC (Intergovernmental Panel on Climate Change) (2007) *Working Group II Contribution to the Intergovernmental Panel on Climate Change, Fourth Assessment Report, Climate Change 2007: Summary for Policymakers*, www.ipcc.ch/SPM6avr07.pdf

McMichael, A. J., D. H. Campbell-Lendrum, C. F. Corvalán, K. L. Ebi, A. Githeko, J. D. Scheraga and A. Woodward (2003) *Climate Change and Human Health – Risks and Responses*, World Health Organization, Geneva

McNeely, J. A. (2006) 'Biofuels: Green energy or grim reaper?', The Green Room, 2 September, BBC, news.bbc.co.uk/2/hi/science/nature/5369284.stm

New Energy for America (2006) *The Apollo Jobs Report: For Good Jobs and Energy Independence*, www.apolloalliance.org/docUploads/ApolloReportper cent5F022404per cent5F122748per cent2Epdf

Nieto, J. (2006) 'Beyond Kyoto Protocol process', Trade Union discussion paper, www.globalunions.org/pdf/ohsewpP_8Bv3.EN.pdf

North American Labor Assembly on Climate Crisis (2007) *Global Warming is Real, Climate Change is Here. We Must Take Action Now*, Cornell Global Labor Institute, New York, www.ilr.cornell.edu/globallaborinstitute/events/climateChangeConference

O'Brien, K., J. Barnett and L. Rosentrater (2007) *Global Environmental Change and Human Security Project*, www.rtcc.org/2007/html/res_edu_gechs.html

Roche, K. and T. Perez (2006) *Our Addiction to Oil is Fueling World Poverty*, 6 April, www.americanprogress.org/issues/2006/04/b1533171.html

Somavía, J. (2007) *Green Jobs Initiative*, ILO, Geneva, www.ilo.org/global/About_the_ILO/Media_and_public_information/Press_releases/lang--en/WCMS_082988

Stern, N., S. Peters, V. Bakhshi, A. Bowen, C. Cameron, S. Catovsky, D. Crane, S. Cruickshank, S. Dietz, N. Edmonson, S.-L. Garbett, L. Hamid, G. Hoffman, D. Ingram, B. Jones, N. Patmore, H. Radcliffe, R. Sathiyarajah, M. Stock, C. Taylor,

T. Vernon, H. Wanjie and D. Zenghelis (2006) *Stern Review of the Economics of Climate Change: Final Report*, HM Treasury, London, Part 11, Chapter 3

UNDP (United Nations Development Programme) (1994) 'New dimensions of human security', in *UNDP Human Development Report*, Oxford University Press, Oxford, pp24–25, www.hdr.undp.org/reports/global/1994/en/

UNDP (2004) *Reducing Disaster Risk: A Challenge for Development*, UNDP, New York, NY, www.un.org/special-rep/ohrlls/ldc/Global-Reports/UNDPpercent20Reducingpercent20Disasterper cent20Risk.pdf

UNDP (2006) *Expanding Access to Modern Energy Services: Replicating, Scaling Up and Mainstreaming at the Local Level*, UNDP, New York, NY

UNISDR (2003) *Living with Risk: A Global Review of Disaster Reduction Initiatives*, Geneva, www.unisdr.org/eng/about_isdr/bd-lwr-2004-eng.htm

USWA (United Steel Workers Association) (2002) *Securing Our Children's World: Our Union and the Environment*, USWA, Cleveland, OH

Warshaw, L. (2007) *Precarious Employment and Child Labour*, www.ilo.org/encyclopedia/?doc&nd=857400057&nh=0&ssect=1

Wisner, M., M. Fordham, I. Kelman, B. R. Johnston, D. Simon, A. Lavell, H. Günter Brauch, U. Oswald Spring, D. Wilches-Chaux, M. Moench and D. Weiner (2006) *A More Secure World: Climate Change and Human Security*, www.radixonline.org/cchs.html

World Commission on Dams (2000) 'People and large dams: Social performance', in *Dams and Development*, www.dams.org//docs/report/wcdch4.pdf

23
What Is a Credible Corporate Response to Climate Change?

Craig Bennett

Introduction

Over the last couple of decades, climate change has moved from an issue that never made it into the corporate boardroom to one that, directly or indirectly, is now influencing many decisions of corporate strategy. But what is a credible corporate response to climate change?

There is, of course, no one model of how companies should respond; but in this chapter I present seven areas of action that are all crucial for any credible corporate strategy. I argue that there are many examples of companies undertaking good, sometimes groundbreaking, work in one or a number of the seven areas. And yet, unless a company's climate strategy addresses all seven areas, its response to climate change will be, at best, inadequate and, at worst, superficial. If the company is weak in just one area, it will inevitably be shaped by the transformational change that will occur this century, rather than help to shape it.

The first five of the seven themes will be familiar to any experienced professional working in corporate responsibility:

1. business operations;
2. products and services;
3. staff training and development;
4. communications;
5. knowledge management and reporting.

Business operations

Many companies start on their sustainability journey by addressing the social and environmental impact of their business operations partly because it is the aspect of the business that their staff may be most familiar with.

Action in this area is always important and in some industrial sectors much can be achieved (and costs reduced) through projects to improve the sustainability of business operations.

Novo Nordisk (the Danish-based pharmaceutical company), for example, has a climate strategy that aims to reduce its dependency upon carbon-based fuel through efficiency gains, energy savings and conversion to renewable energy. Although the strategy covers its global operations, the company has placed particular focus on its Danish production sites, which produce the active pharmaceutical ingredient for its insulin products and, in 2008, represented 85 per cent of the company's total CO_2 emissions. In 2007, Novo Nordisk launched a partnership with Danish energy supplier DONG Energy, in which it identified energy savings and invested the money saved into a new wind energy facility in the North Sea. The company's objective is for all its Danish operations to be supplied entirely by wind energy by 2014 (Novo Nordisk, 2009).

Another example of what can be achieved is provided by the UK telecommunications company British Telecom (BT), which managed to cut its carbon emissions by 58 per cent between 1996 and 2006. Its board has now agreed a new target to cut emissions 80 per cent from 2006 levels by 2016 through a range of initiatives including fleet efficiency improvements, renewable energy supply, carbon foot-printing of all procurement contracts and changes in work practices to save energy and fuel use (CPI, 2008).

In any large company the size of the company's operations means that one seemingly small change to a procedure or procurement contract may have profound implications for carbon emissions across the group and even nationally. BT's 6000 UK sites, for example, are responsible for 0.7 per cent of national electricity use. Its green energy contract, one of the biggest in the world, will result in the company saving the equivalent amount of carbon every year as the electricity used by the populations of Liverpool and Cardiff (CPI, 2008).

The scale of BT's operations is dwarfed, however, by that of Wal-Mart, the world's largest retailer with over 7000 stores in 14 countries. If its 1.9 million workforce lived in a single place, it would be the fifth largest city in the US. By its own estimates, it is the largest private user of electricity in the US. Until recently, it emitted 19.1 million tonnes of carbon dioxide a year (equivalent to 2.8 million households) (Blowfield, 2008).

In 2005, however, Wal-Mart President and CEO Lee Scott launched Sustainability 360, which included a number of commitments – for example, that the company will work towards being supplied by 100 per cent renewable energy. The company is also working with suppliers to reduce packaging and, as a result, hopes to remove 213,000 trucks from the road, saving an estimated

67 million gallons of diesel per year and realizing cost savings of US$11 billion (Scott, 2007), although it is unclear what the timescale is for this.

There is no question that a focus on reducing greenhouse gas emissions from a company's operations can deliver results. It can sometimes be the case, however, that companies put most of their climate effort into this area because it is the most obvious aspect of their business to address, rather than the most important.

I was once made aware of a comprehensive energy saving and paper recycling scheme that had been introduced to the head office of a major British bank while it was simultaneously arranging finance for unsustainable logging operations in Indonesia. It is hard to imagine that the climate 'gain' of turning off a few lights and recycling a bit of paper (while worthwhile) was not far outweighed by the climate change 'cost' of financing the logging operations.

Products and services

A company cannot act 'sustainably' unless it has considered the implications of sustainability for the goods and services its sells, whether food, fossil fuels or financial services. This raises issues about sustainability in the supply chain, but also about the need to develop new, more sustainable products and services.

Just a few years ago, the focus of attention was on how many 'ethical consumers' might exist to purchase 'green products'. The pervading paradigm suggested that as long as enough customers chose the greener products and, by implication, left the less green products on the shelves, then there would be a transformation in the goods and services on offer.

The limitations of green consumerism should have been obvious, however. When 'green products' (however defined) are produced for a niche market of 'green consumers', they are nearly always more expensive than those that are merely produced as cheaply as possible. A more fundamental limit is that even the most ardent, the most caring, the most affluent green consumer will never possess enough knowledge (or time) to buy 'green' all the time. While certification, labelling schemes and even 'carbon footprint' labels exist for some products, they account for only a small proportion of the range of goods that the average consumer buys (Bennett, 2005; Bennett and Burley, 2005).

The average supermarket contains tens of thousands of product lines. The climate and energy issues associated with product supply and use are ever more complex and dynamic. How can we possibly expect consumers to keep abreast of all the latest developments and then have the time to work out for themselves what this means for their shopping basket in a world where people are increasingly time poor? How is a 'green consumer' to express their desire for a particular 'green product' if it does not exist yet?

While 'green consumerism' may have a limited role to play, it seems highly unlikely that it will trigger the transformational change in goods and services that is needed to deliver climate and energy security:

> *The focus needs to be on creating a supportive framework for collective progress, rather than exhorting individuals to go against the grain... It is possible to make sustainable habits and choices easier to take up, by drawing on insights about consumer behaviour and using people's preferences for purchasing shortcuts, and what we call the trend towards 'choice editing'.* (National Consumer Council and Sustainable Development Commission, 2006)

The bulk of responsibility lies, therefore, with companies to ensure that all their products and services are transformed so that they contribute to climate and energy security or are removed from the marketplace when this is not possible. In doing so, there are real opportunities to go beyond narrow concerns about the 'offer' of an individual product and, instead, build a deeper and longer relationship with the consumer in which they develop trust that a certain company or brand is working with them, across all their product lines, to address these issues.

As Bart Becht, chief executive officer (CEO) of the global consumer products company Reckitt Benckiser (which includes household brands such as Cillit Bang, Dettol, Finish, Harpic and Vanish) commented recently:

> *The only way we can have any real impact on climate change is to look not just at the emissions that stem from the manufacturing process, but to examine the total carbon footprint from cradle to grave of what we produce. This means looking at everything, from how we extract the raw materials all the way through to a product's use and disposal. Only by tackling the embedded carbon across our products' entire life cycle can we fully take responsibility and make a meaningful reduction to our global emissions.*
>
> *The reason we at Reckitt Benckiser have adopted this approach is that we directly control only a very small proportion of the emissions and energy use, when compared to the total carbon footprint of a product. For example, we have calculated that our manufacturing and distribution accounts for less than 5 per cent of the total, with raw materials and packaging producing roughly 30 per cent.*
>
> *Perhaps surprisingly, then, our research shows that 60 to 70 per cent of the carbon contribution of Reckitt Benckiser products rests with consumers and how they use and dispose of our products. So despite our achievements in reducing our carbon emissions from manufacturing by 30 per cent since 2000, there is much more we can all do. This is not about passing the buck onto our customers; it is about working with them.* (Becht, 2008)

In 2007, Reckitt Benckiser launched its Carbon 20 programme in which it committed to reduce its 'total carbon footprint' by 20 per cent by 2020 (on a base of circa 15 million tonnes of CO_2 equivalent per annum). If Reckitt

Benckiser achieves its targets, it will have removed the carbon impact equivalent of nearly 1 million medium-sized family cars (Reckitt Benckiser, 2007).

It is worth noting the political significance of Reckitt Benckiser's commitment. In 2008, the European Union finally adopted a target to reduce its greenhouse gas emissions by 20 per cent by 2020, but only after long and protracted negotiations and resistance from many member states. The vast majority of developed countries are yet to adopt similar mid-term commitments. This begs the question: if one of the world's largest consumer products companies thinks such reductions are possible even when their commitments extend into those areas outside their direct control, why are so many of the world's politicians so hesitant to make similar commitments?

In some circumstances, it may be that a particular type of product or service has no place in a carbon-constrained world. In these cases, it will be necessary for a company to withdraw the product from the marketplace. An excellent, if almost comical, example of this so-called 'choice editing' is the decision that some UK retailers made in 2007 to stop selling patio heaters. Patio heaters burn propane gas or use electricity to provide outdoor heating and became popular with domestic consumers at the start of the decade.

The problem is that they are incredibly wasteful of energy. The UK-based Energy Saving Trust has estimated that a propane patio heater with a heat output of 12.5kW will produce around 34.9kg of CO_2 before the fuel runs out (after approximately 13 hours). This is equivalent to the energy required to produce approximately 5200 cups of tea (or 400 cups for every hour of operation) (Energy Saving Trust, 2007).

The irony of this statistic is that, in an attempt to save energy, the UK government has paid for television advertising to persuade its citizens to avoid overfilling their kettles. A diligent individual could do this for a whole year and the total energy saving would be wiped out if they then sat under a patio heater for just one hour.

A modelling exercise by the UK government's Market Transformation Programme on the energy use of the 630,000 UK domestic patio heaters calculated that they could produce a total of 140,000 tonnes of carbon dioxide per annum. This is roughly equivalent to the carbon emissions from all the homes in a small city such as Bath (Market Transformation Programme, 2007).

This level of emissions associated with just one product, and very clearly a non-essential product at that, prompted Wyevale Country Gardens, the UK's leading garden centre company, to announce in 2007 that it would stop selling patio heaters. In a statement, Wyevale (2007) said:

> *Out of all the products we sell, patio heaters are the least defensible. When we were asked how we could sell something which directly heated the atmosphere, there wasn't much we could say.*

It did not take long for some of its rivals to follow suit. The individual product line was profitable; but these companies could not defend continuing to stock such a wasteful product given their wider corporate responsibility strategies.

'Choice editing' need not be limited to companies that sell physical products. The pioneering work in the financial sector is now about how banks can mainstream sustainability criteria across all their investments and financial services, not just their socially responsible investment (SRI) funds. Some individuals and companies in the advertising industry are starting to wonder whether they should decline contracts to market the most polluting products in the same way that many advertising companies shunned tobacco money during the 1970s.

As Alan Knight, a member of the UK Sustainable Development Commission, has said: 'Customers do not want to be overwhelmed with choice and are happy that choice is limited to only greener options' (Knight, 2007). Every company should be looking at how it can address climate and energy security through the products and services it offers or, just as importantly, does not offer.

Staff training, development and rewards

When companies are serious about improving staff performance or changing practices in a given area of the business, they will normally make a big investment in staff training and development to do this. It is normal nowadays for companies to send all staff on health and safety training, management training, project management training, and so on. Similarly, companies that are serious about sustainability need to build capacity across the whole staff body on the issue. A company will not have 'embedded' sustainability if all the expertise is located within one small department or team.

Fortunately, leadership development programmes in sustainability now exist. The Prince of Wales's Business and Environment Programme (BEP), for example, was established in 1994 and now runs international leadership seminars on sustainability issues and strategy in six locations around the world. Each seminar provides a deep immersion for senior executives in key challenges of sustainability, but then helps them to integrate social and environmental solutions within decision-making processes while still remaining profitable.

BEP is the flagship programme of the University of Cambridge Programme for Sustainability Leadership (CPSL), an institution within the university that works with business, government and civil society to build leaders' capacity to meet the needs of society and address critical global challenges. Other CPSL programmes include the Climate Leadership Programme, which helps senior executives to seize the business advantage inherent in addressing global climate change, and Financing the Future, which seeks to equip senior leaders in the finance sector with an understanding of key climate change impacts, trends and responses, and to explore the risks and opportunities that these present for financial-sector organizations and their markets.

As more and more business leaders have gained an understanding of sustainability issues, so many companies have adopted an approach to business that recognizes the role of social and environmental issues in enhancing commercial performance. But if this strategy is to succeed, employees need to see the sustainable development objectives clearly reflected in their personal targets and rewards.

Performance appraisal must take into account the contribution of individuals and teams to longer-term social and environmental goals as well as short-term financial objectives. If all the bonuses for a procurement team are dependent upon achieving lowest cost contracts, for example, it will be no surprise if they display little interest in the energy performance and carbon intensity of the supplier base.

I know of one CEO of a FTSE 100 retail company that has a quantified target for carbon reduction as one of the seven objectives that will be used to assess his next bonus. He has since made sure that this is reflected in the objectives of his direct reports and it is expected that the objective will cascade down through the organization increasing the likelihood that it is met.

Knowledge management and reporting

If a company is in any way serious about reducing its carbon footprint, it needs to develop sound metrics for measuring and reporting progress on its carbon reduction strategy and emissions. As Lord Turner, chairman of the UK's Financial Services Authority (and briefly chairman of the UK's Committee on Climate Change), has said: 'The first step towards managing carbon emissions is to measure them because in business what gets measured gets managed' (Carbon Disclosure Project, 2009).

The reality is that carbon emissions are now a financially material commodity and will become more so. They have an economic and financial value to business, investors and suppliers. They need to be properly defined, measured, accounted for, audited and reported in the same way as other physical commodities and financial instruments (Aldersgate Group, 2007).

The Global Reporting Initiative, the World Business Council on Sustainable Development (WBCSD), the UK government's UK Department for Environment, Food and Rural Affairs (Defra) and the Carbon Trust, to name but a few, have all produced guidelines and definitions for reporting on carbon emissions. The absence of comparable standardized measures means, however, that companies that reduce their carbon emissions are unable to demonstrate their success in relation to their competitors. Where individual companies have changed their carbon reporting methodologies over time, it also means that comparisons cannot be made year on year.

In my view, it is only a matter of time (and probably not long) before governments introduce a common and legally binding methodology for how companies should report their carbon emissions. The companies that are best placed to implement will be those that have already gained experience by participating in one of the voluntary schemes.

Investors and other stakeholders are increasingly interested in the broader management context of carbon emissions, however, and not just the hard data. The Carbon Disclosure Project (CDP) is an independent not-for-profit organization that holds the largest database of corporate climate change information in the world. The data is obtained from responses to CDP's annual information requests, issued on behalf of institutional investors, purchasing organizations and government bodies to more than 3700 corporations across the globe (Carbon Disclosure Project, 2009).

Its information requests cover four main areas:

1 management's views on the risks and opportunities that climate change presents to the business;
2 greenhouse gas emissions accounting;
3 management's strategy to reduce emissions/minimize risk and capitalize on opportunity; and
4 corporate governance with regards to climate change.

Companies that are unable to provide comprehensive and credible answers to questions in these areas will increasingly be seen as lacking in key management competencies and strategic analysis.

Internal and external communications

The development and implementation of sustainability strategies requires a holistic and joined-up approach to ensure that action in one area does not compromise progress in another. It also requires the understanding, 'buy-in' and support of management, employees, customers, investors, suppliers and customers alike.

For these reasons, good internal and external communications can be crucial to ensure the successful development, implementation and long-term resilience of a company's sustainability strategy. Communications play a particularly important role in communicating the business case for action.

The British retailer Marks & Spencer (M&S) offers what is widely regarded amongst corporate sustainability professionals as one of the best examples of how good corporate communications can help to deliver sustainability and corporate objectives. During the 1990s and early part of this decade, the company was finding it hard to differentiate itself in a positive way from many of its competitors, particularly the larger supermarket chains (which had expanded into clothing and other non-food items once considered the preserve of M&S). M&S needed something that would enable the company to reassert its traditional strengths to its customer base but in a fresh way.

In January 2007, and with a very high-profile advertising blitz, the company launched *Plan A*, which represented the company's response to 'the most important social, environmental and ethical challenges we face'. It included '100 Plan A commitments' that 'set sustainability targets for 2012'. The launch text said: 'We're doing this because it's what you want us to do. It's also the right thing to do. We're calling it Plan A because we believe it's now the only way to do business. There is no Plan B' (Marks & Spencer, 2009).

The 100 Plan A commitments were categorized into five headings: climate change; waste; sustainable raw materials; fair partner; and health. The climate change category is introduced with the following preface:

> *We aim to make all our UK and Irish operations carbon neutral by 2012. We'll maximize our use of renewable energy and only use offsetting as a last resort. And we'll be helping our customers and suppliers to cut their carbon emissions too.*

In total, there are 29 commitments on climate change, including:

> 1 *Carbon neutral: aiming to make all our UK and Republic of Ireland operations (stores, offices, warehouses, business travel and logistics) carbon neutral.*
> 2 *Energy efficiency (stores): reducing the amount of energy we use in our stores by 25 per cent per square foot of floor space...*
> 7 *Green electricity: sourcing or generating 100 per cent 'green' (renewable) electricity for M&S stores, offices and distribution centres in the UK and Republic of Ireland...*
> 10 *On-site renewables: having 20 per cent on-site energy generation from renewables in all new builds where practicable.* (Marks & Spencer, 2009)

Plan A was promoted heavily through TV and print advertising, in store displays and bold signage on delivery vehicles. Most adverts or displays focused on just one of the 100 Plan A commitments and gave the impression that it has been selected by random from the longer list.

The campaign was very successful because it combined passionate simplicity ('There is no Plan B') with a level of complexity and detail (achieved through the individual commitments) in a way that conveyed decisiveness, substance and credibility.

I know of many corporate sustainability professionals that are jealous of what M&S achieved with Plan A. It might be that other companies had similar or equivalent sustainability commitments in place; but none of them managed to convey them in such an engaging, yet comprehensive manner, and in a way that has helped to deliver on much broader corporate objectives. In other words, it was the communications element to Plan A that made it successful on both sustainability and commercial criteria.

There are, of course, plenty of companies that have done a poor job in communicating their work on climate change. I know of corporate responsibility professionals that have put good climate change strategies in place but because they have failed to communicate them in a positive and engaging way to colleagues, customers, shareholders and other stakeholders, their work has been called into question.

Just as common, unfortunately, is the propensity for corporate communication departments to do a 'PR job' on climate change and overstretch a company's performance. 'Greenwash', as it has become commonly known, is easy to spot and, in my experience, does far more harm than good to a company's credibility in the debate.

A company's communications on climate change should seek to tell a simple and honest story: where they have come from on the issue, where they are now, what their future strategy is and why. If it can do this in an interesting and engaging way, using targets and data where appropriate to avoid bland assertions and 'wish lists', and if it can share challenges just as openly as it shares successes, then its efforts will be recognized and appreciated by the vast majority of stakeholders.

These first five areas of action will be familiar to many people who follow how companies are responding to climate change. Sound performance is needed in each area if a company is to avoid being exposed to reputational risk or being caught out by the rapidly changing policy context.

There is a limit, however, to the level of change that can be brought inside or outside a company by action in these five areas. Even if the majority of large companies were performing well in each of these areas, the most that would be achieved would be incremental change. And, yet, the scale of the climate change challenge requires nothing short of a transformational change in how we manage the global economy.

It is for this reason that I argue that for a company to have a truly comprehensive and credible response to climate change, its strategy must also ensure that it is taking action in two additional areas: business model and strategy, and interaction with public policy.

Business model and strategy

The 21st century will present extraordinary challenges for many sectors that are key components of today's economy.

The companies that are able to benefit rather than suffer from these challenges will be those that are prepared to evolve their business model to the transforming economy, rather than remain static and/or seek to defend and maintain their old business model as it becomes increasingly irrelevant.

An example of a company that went through a genuine transformational change by modifying its business model and strategy to changing circumstances is the world's largest manufacturer of mobile phones, Nokia.

When Nokia started in 1865, it made paper. During its first century, it became a major industrial force with growing interests in rubber and cables, as well as pulp and forestry. During the 1960s and 1970s, the company started to play a pioneering role in the early evolution of mobile communications and in the 1980s it began to commercialize its in-car phones for use on Finnish highways.

The transformational change that occurred to Nokia in the 1990s, however, can be traced back to a decision made by Jorma Ollila who became its president and CEO in 1992. He made the crucial and, at the time, extremely bold strategic decision to focus on telecommunications and move Nokia out of its traditional sectors. During the 1990s, the rubber, cable and consumer electronics divisions were gradually sold as the company developed a strong focus on mobile communications and, with that focus, the fortunes of the company changed dramatically.

By 2005, just 13 years after Ollila's decision to refocus the company, Nokia has sold its billionth handset. Two years later, Nokia became the fifth most valued brand in the world.

Nokia's extraordinary success occurred because it was willing and able to redefine itself to reflect technological breakthroughs and changing societal behaviours. In so doing, it has also driven much of the technological and behavioural change associated with mobile communications over the last two decades. In some respects, it would be true to say that the company changed and, in turn, it changed the world.

Although Nokia's transformation was not driven by issues relating to sustainable development, it provides an excellent example of the sort of opportunities that will exist over the coming decades for those companies that are ready and willing to redefine themselves as the world changes around them – and how they, in turn, can drive deeper system change.

The more companies that are able to redefine themselves to meet the needs of a low-carbon economy, the faster that low-carbon economy will be achieved. Changing a company's long-term business model and strategy to reflect this may represent the single most important action it could take in response to climate change, both from a commercial and sustainability perspective.

Interaction with public policy

A company's impact upon sustainability, however, goes far beyond its role merely as a trading body.

As previously noted, the sustainability challenge requires nothing short of a transformational change to the way in which we manage our economy. Incremental change (i.e. that which can be delivered by companies acting alone) will fail to address the scale and urgency of problems such as climate change, breakdown of ecosystem services, water scarcity and growing inequality.

Transformational change will only be possible through government interventions that are guided by a clear, consistent and strategic approach to policy-making – and where long-term regulatory frameworks are put in place to correct market failure and to change the terms of trade for resource consumption. Governments will only be able to introduce such bold and long-term policy measures if they are given support and political space by the business community to do so, and so the role that companies can play as progressive 'corporate citizens' – urging governments to introduce transformational policies – becomes critical.

The corporate sector's history in public policy has not been good. The public image of a 'corporate lobbyist' is of someone working to protect a narrow vested interest, often against the interests of society as a whole. There are plenty of famous examples of companies, or groups of companies spending millions of dollars in opposition to progressive public policy, whether it is tobacco companies in the 1960s and 1970s questioning the link between smoking and cancer, aerosol companies lobbying against the banning of chlorofluorocarbons (CFCs) in the 1980s, or oil companies clubbing together as the Global Climate Coalition to question the science of climate change in the 1990s.

The widespread perception of 'corporate influence' as a malign force (as described in Monbiot, 2000, and Korten, 1996) is not, however, just derived from these famous examples. The way in which mainstream business associations are structured has historically resulted in them often adopting 'lowest common denominator' positions – where a large group of companies ends up adopting the position of the least progressive member as a lazy way of securing consensus. A business association might then assert the agreed position as if it was held passionately by its members as a whole, even though it might more accurately reflect the position of just a handful of the least progressive members.

For many years, this discrepancy was not of material concern to companies, especially on issues to do with environmental policy.

This was clearly illustrated by a conversation I had, in 2003, with a very senior executive at one of the UK's largest companies. The company had a fairly progressive position on climate change and I asked him what he was doing to challenge the position of the Confederation of British Industry (CBI) – which remained sceptical of the need for action on climate change. He laughed about the CBI's position, told me 'not to worry' and explained that they only bothered to send junior staff to the CBI policy meetings.

His company was one of the CBI's biggest contributors and he had completely missed the point of my question. I had hoped to ascertain what his enlightened company might have been able to do to influence the CBI's position on climate change, not to hear that it was going to sit back and let it say whatever it wanted while his company filled its coffers.

This sort of lazy indifference mattered. By 2004, under the charismatic leadership of Sir Digby Jones, the CBI was becoming increasingly hostile to environmental regulations and was repeatedly warning that they would cause UK-based companies to relocate (see, for example, CBI, 2004). The CBI had little evidence to substantiate this assertion. When, in January 2005, the House of Commons Environmental Audit Committee asked Sir Digby for an example of a company that had moved overseas as a result of environmental pressures, he said there were 'nil' (House of Commons Environmental Audit Committee, 2005).

Just a couple of months later, however, the CBI published its business agenda to coincide with the 2005 UK general election and called on the future government to 'set emissions reduction targets ... that are sensitive to the need not to simply drive business overseas' (CBI, 2005).

The CBI's behaviour infuriated non-governmental organization (NGOs) and even led Friends of the Earth-UK to launch a campaign report on the CBI (see McRae et al, 2005). But the business association's tactics were also causing disquiet amongst some of its largest members who were starting to feel uncomfortable at how business was repeatedly being portrayed as sceptical of the need for action on climate change. Their concern was not so much about how they might be perceived by members of the public, but more about how UK plc might miss out on the opportunities to develop new low-carbon technologies if the government failed to introduce relevant policies because it was running scared of a hostile business lobby.

As a result, in June 2005, the UK Corporate Leaders Group on Climate Change (CLG) was convened, initially by just 13 companies, including Johnson Matthey, F&C Asset Management, BAA, Standard Chartered Bank, John Lewis Partnership, Shell and Sun Microsystems. Its aim was to bring together business leaders to advocate the development of new and longer-term policies for tackling climate change. Its first output was an open letter to Prime Minister Tony Blair in May 2005, published immediately in advance of the G8 Gleneagles Summit. It noted that enabling a low-carbon future should be 'a strategic business objective for the UK' and stated:

> *At present, we believe that the private sector and governments are caught in a 'Catch 22' situation with regard to tackling climate change. Governments tend to feel limited in their ability to introduce new policies for reducing emissions because they fear business resistance, while companies are unable to take their investments in low carbon solutions to scale because of lack of long-term policies.* (Corporate Leaders Group, 2005)

In order to help break this impasse, the CLG offered to work in partnership with the UK government in order to:

- support the development of a world-leading climate change policy framework capable of enabling a step change in private-sector investment in low-carbon technology in the UK;
- significantly increase support for action on climate change from UK businesses, the public and other governments and businesses internationally;
- dramatically scale up investment in low-carbon technologies and processes by our companies and others in response to new policy (Corporate Leaders Group, 2005).

The letter was followed by a meeting between the CEOs of UK CLG member companies and Prime Minister Tony Blair and the modus operandi of the CLG was born.

In 2006, the group sent a second letter to the prime minister in which they called on the UK government to set tougher targets on carbon emissions under the European Union Emissions Trading Scheme (ETS). Towards the end of the year, a different group of pan-European companies came together to form the EU Corporate Leaders Group on Climate Change (CLG). They sent an open letter to President Barroso in which they called on him to introduce the long-term policy frameworks that were needed at the EU level to tackle climate change. They welcomed the development of the ETS, but noted that it would not be enough and that other policy interventions would be required. Crucially, they argued that the EU's overall competitiveness need not be harmed by tackling climate change and, on the contrary, there was a competitive advantage to be gained in growing a global market for low-carbon technologies.

Since then, the UK and EU corporate leaders groups have made a number of very significant interventions in the UK, EU and international policy debates on climate change. In 2007, they came together to publish a communiqué to governments gathering at the UN Climate Change Conference in Bali. The Bali Communiqué was endorsed by over 170 companies from around the world. It called for:

- a comprehensive, legally binding United Nations framework to tackle climate change;
- emission reduction targets to be guided primarily by science;
- those countries that have already industrialized to make the greatest effort;
- world leaders to seize the window of opportunity and agree a work plan of negotiations to ensure an agreement can come into force post-2012.

The communiqué appeared in a full-colour, centre-spread advert in the global edition of the *Financial Times* on Friday, 30 November 2007, just before the Bali conference and has been credited by some climate negotiators as having played a crucial role in persuading the US to change its position at the conference.

Just a decade ago, it would have seemed inconceivable to many that business groupings would be formed to advocate bold regulatory interventions from governments in support of sustainability.

And yet, this seventh area of how companies should be engaging with sustainability issues could prove to be the most important since it only through progressive interactions in the public policy debate that companies can help to drive the transformational change that is necessary to make all companies truly sustainable.

References

Aldersgate Group (2007) *Carbon Costs: Corporate Carbon Accounting and Reporting*, Aldersgate Group, London

Becht, B. (2008) 'Economy offers chance to act on climate change', *The Daily Telegraph*, London, 30 December, p24

Bennett, C. (2005) 'Protecting the environment through global regulation', in A. Lent (ed) *Just World*, The Fabian Society and Zed Books, London

Bennett, C. and H. Burley (2005) 'Corporate accountability: An NGO perspective', in S. Tully (ed) *Research Handbook on Corporate Legal Responsibility*, Edward Elgar, Cheltenham, UK

Blowfield, M. (2008) *Wal-Mart – The Retailer of the Future? Cambridge Sustainability Network: Case Study*, Cambridge Programme for Sustainability Leadership, Cambridge, UK

Carbon Disclosure Project (2009) *About Us*, www.cdproject.net

CBI (Confederation of British Industry) (2004) *UK Environmental Regulations: The UK – A Place To Do Business*, CBI, London

CBI (2005) *The CBI Business Agenda*, CBI, London

Corporate Leaders Group (2005) Open letter to Prime Minister Tony Blair, 27 May, www.hsbc.com/1/PA_1_1_S5/content/assets/newsroom/primeministerletter.pdf

CPI (Cambridge Programme for Industry) (2008) *Finding the Way: Examples of Responsible Business,* CPI, Cambridge, UK

Energy Saving Trust (2007) *Housing and Buildings*, www.energysavingtrust.org.uk

House of Commons Environmental Audit Committee (2005) *Oral Evidence*, 19 January 2005

Knight, A. (2007) *The Bigger Picture*, Wyevale Country Gardens, p4

Korten, D. (1996) *When Corporations Rule the World*, Berrett-Koehler, San Francisco

Market Transformation Programme (2007) *Approved Briefing Notes*, www.mtprog.com

Marks & Spencer (2009) *Plan A*, www.marksandspencer.com

McRae, S., C. Bennett and H. Burley (2005) *Easy Listening: Why the Government Needs a New Playlist Featuring More than the CBI's Old Favourites – A Research Summary*, Friends of the Earth, London

Monbiot, G. (2000) *Captive State: The Corporate Takeover of Britain*, Macmillan, London

National Consumer Council and Sustainable Development Commission (2006) *I Will if You Will – Towards Sustainable Consumption*, NCC/SDC, London

Novo Nordisk (2009) *Novo Nordisk Annual Report: Financial, Social and Environmental Performance*, Novo Nordisk, Denmark

Reckitt Benckiser (2007) *Reckitt Benckiser Targets a New Standard in Carbon Reduction*, Press releases, 1 November, www.reckittbenckiser.com

Scott, L. (2007) 'Sustainability 360°: Doing good, better, together', 2007 London Lecture, The Prince of Wales's Business and Environment Programme, Cambridge Programme for Industry, Cambridge, UK

Wyevale Garden Centre (2007) *People, Plants and the Planet*, Wyevale Country Gardens

24
Taking Action: Financing a Climate Change Technology Revolution

Andrew Higham[1]

Introduction

This chapter examines the financial dimension of climate change – how much money is needed, where the money might come from and what policy options are available to fill the finance gap between what is currently available and what is required. It also examines the potential role of the Climate Change Convention in the financing of climate change technologies.

The chapter reports on a programme of research and policy advice commissioned in Bali at the 13th Meeting of the Conference of Parties (COP 13) in December 2007. The Bali Action Plan (Decision 1 and 3/CP.13), which provides the road map for a post-2012 climate change agreement, places finance at the heart of a successful negotiated outcome. In particular, it emphasizes the need for adequate resources and the redistribution of financial resources from developed to developing countries.

The plan (paragraph 1(e) Decision 3/CP.13) calls for 'enhanced action on the provision of financial resources and investment to support action on mitigation and adaptation and technology cooperation, including ... mobilization of public- and private-sector funding and investment, including facilitation of carbon-friendly investment choices'. The plan directly links developing countries' emissions reduction actions with the scale of financial support provided by developed countries.

It is not possible to rely upon developed country emissions reduction commitments – even if it were possible to totally eliminate developed country emissions. Only with a combination of deep cuts in developed country emissions (in the order of 25 to 40 per cent below 1990 levels by 2020, greater than 50 per cent by 2050, with developed country emission reductions in the order of 80 to 90 per cent by then), with developing country actions to limit emissions by 15 to 30 per cent below business as usual over the same time period, will it be possible to avert dangerous climate change and keep global warming to around 2°C (den Elzen and Höhne, 2008; Metz, 2008).

The Bali Action Plan (Decision 1/CP.13 paragraph 1(a) (i) and (ii))[2] committed developing country parties to negotiate nationally appropriate mitigation actions (NAMAs) that are measurable, reportable and verifiable (MRV), and developed country parties to quantified emission limitation and reduction objectives, commitments and actions that are MRV.

For developed countries the MRV requirement also includes commitments to technology, financing and capacity-building support for developing countries. If developed countries fail to provide adequate financial support, there can be no chance that developing countries will commit to limit their emissions; consequently, there can be no chance of preventing dangerous climate change.

According to the secretariat of the United Nations Framework Convention on Climate Change (UNFCCC), additional investments in climate technologies of US$441.5 billion to $1436 billion will be required, on average, each year from 2010 to 2050 (IEA, 2008b; Higham, 2009). While investment needs by sector change over time, on average the investment needs double after 2030 compared with investment needs from 2010 to 2030. Current annual investment in climate change mitigation technologies is estimated at US$70 billion to $155 billion (UNFCCC, 2009). A total of 40 to 60 per cent of the increase, an additional US$177 billion to $862 billion per annum, is projected to be needed in developing countries, much of which will require support and financial transfers from developed countries. This reflects the scale of the emissions reduction potential estimated to be available in developing countries. In summary, upfront investment in climate change mitigation technologies will need to increase fivefold to tenfold.

The type of finance will be important depending upon how the incremental costs are defined. If the fuel savings and co-benefits of investments are taken into account, then loans might become more feasible depending upon the cost of abatement and the prevalent carbon price, and assuming the absence of some defining barrier to financing technology in developing countries.

Much of the upfront investment will be returned over time to the owners and operators of the investments and technologies in the form of reduced fuel costs or greater efficiency or productivity. Full savings from upfront investment costs at a discount rate of 10 per cent reduces the overall cost of achieving a 500 to 550 part per million (ppm) carbon dioxide equivalent emissions stabilization scenario by 2050 to 800 billion, or just over 50 billion per year on average.

It will also be necessary to invest heavily in adaptation to the unavoidable impacts of climate change in all nations, but particularly in the more vulnerable developing countries. These investments are likely to be financed largely by the public sector as many will not derive a financial return or are sub-economic (UNFCCC, 2008a). The World Bank estimated that adaptation will cost US$10 billion to $40 billion and Oxfam International estimated it will be more than US$50 billion annually (Raworth, 2007). In 2007 the United Nations Development Programme (UNDP) projected the annual adaptation investment needs as US$86 billion by 2015. Christian Aid estimates adaptation costs at US$100 billion per year in developing countries (Flåm and Skjærseth, 2009, Table 1, p110). The UNFCCC estimates the additional investment and financial flows in 2030 as US$49 billion to $171 billion globally, of which US$28 billion to $67 billion is for developing countries (UNFCCC, 2007, Table IX–65, p177). These estimates of investment needs are in stark contrast with current adaptation investments in developing countries of US$0.4 billion to $0.6 billion per year (Haites et al, 2009). Therefore, investment in climate change adaptation technologies will need to increase by at least 100-fold.

Negotiations under the Climate Change Convention have identified several feasible opportunities to raise new financial resources that do not rely upon voluntary contributions from nations to a total value of US$20 billion per annum (UNFCCC, 2007, 2008a, 2009). The options include auctioning national emission rights (known as 'assigned amounts' in the lexicon of the convention), placing levies on international shipping, air freight and travel, and taxing the transactions that occur through the flexibility mechanisms of the convention – the Clean Development Mechanism (CDM), joint implementation (JI) and international emissions trading. Whether or not governments will agree to introduce these new financing options is uncertain. Clearly, it will not be possible to achieve a transition to a low-carbon and climate change-ready world with just an additional US$20 billion per annum. However, the less money that is raised internationally, the more that will be required through voluntary contributions from governments and the more the convention will be dependent upon national policies and measures.

Financial needs by stage of technological maturity

The concept of 'stages of technological maturity' provides a measure of the progress of a technology from the research laboratory to a widely available, commercially viable product. These stages describe a technology's progress through successive phases of technological innovation. Studies have shown that while significant abatement can be achieved through existing cost-effective technologies, new technologies will be needed and they will need to become much cheaper relative to the incumbent technologies widely used in society. The innovation process is shown in Figure 24.1.

An extensive literature suggests that the interconnections between markets, research institutions and government, united in a 'national innovation system', are essential for the success of technological development. Based on its experience with new technologies in the UK, the Carbon Trust indicates that innovation requires simultaneous progress with corporate evolution, market strategy and regulatory development because they interact with and 'feed back' on each other (OECD, 1997).

While most commentators agree that linear progression in technological innovation is rare, a practical non-linear model of technology innovation has not emerged. The literature includes many different characterizations of this process and the associated stages of technological maturity. Although an oversimplification, for convenience, this chapter characterizes technology innovation as proceeding linearly from research and development, to demonstration, to deployment, to diffusion and to commercially mature.

The stages of technological maturity adopted for the analysis of mitigation technologies – research and development, demonstration, deployment, diffusion and commercially mature – are shown in Figure 24.2. Each stage is defined by barriers that need to be overcome to develop a commercially mature technology. These barriers help to identify the financing vehicles appropriate to each stage. The financing needs of climate change mitigation technologies are determined through a detailed assessment of 165 mitigation technologies across all sectors of the economy that have been identified in 25 global technology studies, financing mechanisms and technology development programmes (UNFCCC, 2008b, 2009).

Research and development means that while the basic science is understood, the technology is at the stage of conceptual design or testing at the laboratory or at the bench scale. The unique barriers that it faces relate to the proof of concept and to technological challenges. Research and development typically occurs in only a few institutions globally for a given technology.

Demonstration involves full-scale implementation of a limited number of installations by a small number of companies or research facilities. Demonstrations provide information on the capital and operating costs and performance of the technology at full scale. This information is used to improve

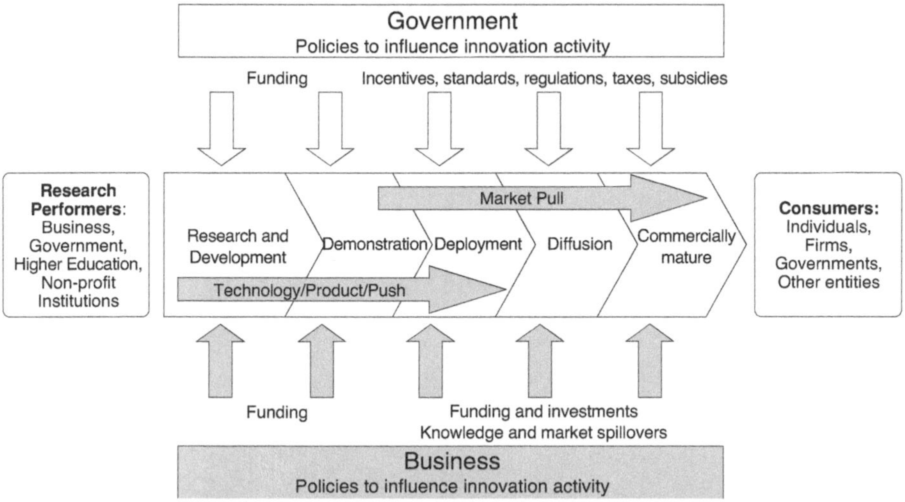

Source: Halsnæs et al (2007, Figure 2.3, p157)

Figure 24.1 *The innovation process*

the cost, performance or other characteristics to make the technology attractive to potential consumers. Finance is often shared between the public and private sector in carefully designed partnerships and contracts; yet it is often difficult for technology developers to make the transition through the demonstration phase – the costs involved increase dramatically and because both public and private finance is difficult to obtain, this stage is often termed the 'valley of death' (US SBA, 1994; Branscomb and Auerswald, 2002; Auerswald and Branscomb, 2003; Murphy and Edwards, 2003; Williams, 2004; Auerswald et al, 2005).

A technology at the deployment stage is well understood and is available for selected commercial applications, but is more costly than the established technology, even taking into account a price for greenhouse gas emissions or equivalent policy. The buyers must pay a premium price, owners must accept a loss on each sale, or governments must provide financial or other incentives for the technology. The experience gained from additional sales usually enables the cost of the technology to be reduced.

At the diffusion stage the technology is competitive with the established technology if a price of greenhouse gas emissions or equivalent policy is taken into account. However, the technology may still face barriers relating to the economic environment, social acceptance or institutional arrangements, such

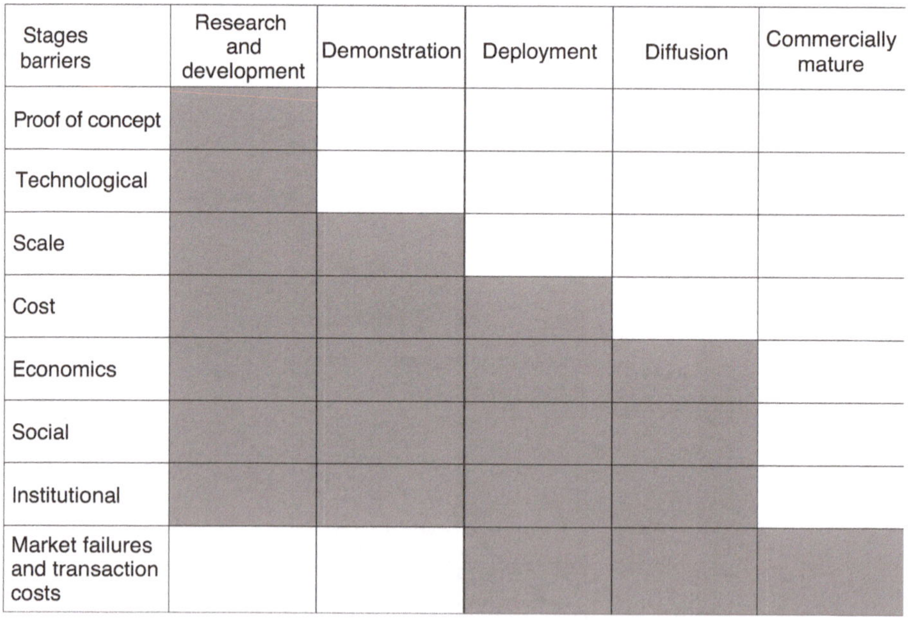

Source: UNFCCC (2008c)

Figure 24.2 *Stages of technological maturity and barriers*

as access to the grid for the sale of electricity generated or the adoption of appropriate safety standards.

A commercially mature technology is competitive with the established technology even if the price of greenhouse gas emissions is not considered, but may need to overcome market failures and specific transaction costs. The market failures faced by energy-efficiency technologies are a classic example. Existing subsidies to fossil fuels and other greenhouse gas emitting technologies are another example.

The unit cost of technology typically falls as the cumulative sales increase, as shown by the experience curve in Figure 24.3. The unit cost of a mitigation technology declines as the cumulative sales increase and it moves from the demonstration to the deployment, diffusion and commercially mature stages. The effectiveness of policies to stimulate adoption of a technology and the success of the innovation effort determine how quickly it moves from one stage to the next. With increased volume, cost reductions can come from the manufacturing process, the distribution system or support services, as well as the technology itself.

The specific barriers to financing climate change technologies are summarized in Table 24.1.

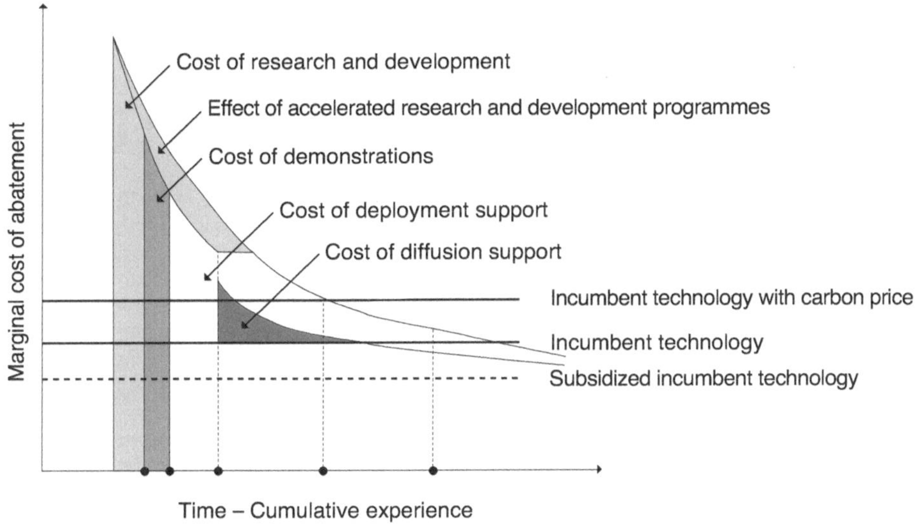

Source: UNFCCC (2008c)

Figure 24.3 *Experience curve of technology innovation*

Table 24.1 *Specific financing barriers related to stage of technological maturity*

Stage of technological maturity	*Category of barriers*	*Financing barriers*	
		Public finance	*Private finance*
Research and development	Proof of concept	• Other political priorities for public finance • Unclear results of fundamental research • Unclear results of education and training	• Insufficient rate of return • Spill-over effects prevent private financiers from capturing benefits of investment
Research and development	Technical	• Other political priorities for government budgets and public finance	• Lack of good technical information, resulting in high-risk profiles • Spill-over effects prevent private financiers from capturing benefits of investment
Research and development; demonstration	Scale	• Relatively high costs to scale up from prototype	• Lack of technological track record, resulting in high-risk profiles
Research and development; demonstration; deployment	Costs	• High costs to reach significant deployment	• Lack of policy to overcome costs, leading to low internal rate of return (IRR)
Research and development; demonstration; deployment; diffusion	Economic	• Unwillingness to interfere in the market, especially when drastic changes harm vested interests • Inflexibility of tax policy	• Energy pricing and subsidies; lack of, or insufficient, carbon price • High upfront capital costs • Lack of valuation of co-benefits, leading to low IRR • Requirement of large parallel infrastructure, leading to high upfront costs
Research and development; demonstration; deployment; diffusion	Social	• Vested interests in social/ consumer preferences • Underinvestment in education and training	• Lack of a consumer or user market • Split incentives (principal–agent problem) • Lack of labour skills
Research and development; demonstration; deployment; diffusion	Institutional	• Vested interests in institutional settings • Public finance policy failures	• Lack of regulatory framework • Absence of international standards • Technology lock-in • Lack of match between export credit agency (ECA) conditions and local finance conditions on environmentally sound technologies

Table 24.1 *(Continued)*

Stage of technological maturity	*Category of barriers*	*Financing barriers*	
		Public finance	*Private finance*
Commercially mature			
Market failures and transaction costs			
• Lack of recognition of public role in resolving market failures and transaction costs • Vested interests in bureaucracies • Inefficient regulatory environment and bureaucracy • Lack of risk assessment and management tools specific to environmentally sound technologies (ESTs) • Lack of appropriate financial packages • Lack of awareness and information • Imperfect markets • Technology market failure			
IEA (2007); New Energy Finance (2008); UNEP (2002);UNEP (2007)			

Source: adapted from UNFCCC (2009); Haites et al (2009)

Discussion of financing for technology development is hampered by the lack of an agreed definition and a paucity of data. The definition of the financing resources needed for technology development adopted for this chapter is illustrated by the shaded area in Figure 24.3. The financing resources needed for technology development are the:

- full cost of activities during the research and development and demonstration stages; plus the
- additional cost of the new technology during the deployment and diffusion stages.

The other important cost is the public investment in programmes that develop facilitative policies and eliminate the barriers that prevent commercially mature technologies such as energy efficiency measures from widespread deployment. This cost is not factored below; but current proposals for such programmes total to US$5 billion to $10 billion per annum globally and in developing countries.

Table 24.2 summarizes the financing needs by stage of technological maturity.

Public and private investment shares

Financing for climate change mitigation technologies is dominated by private sources of finance (approximately 80 per cent, or US$112 billion to $184 billion in 2007); but those private investments are heavily dependent upon public investments and policy frameworks that provide the necessary incentives and market conditions that enable investment (UNFCCC, 2008b; New Energy Finance and UNEP, 2008). Therefore, the public share of investment necessary to address climate change will depend upon the extent to which

Table 24.2 *Estimates of additional financing needed for the development of mitigation technologies (US$ billion per year)*

	R&D (total spending: US$ billion)	Demonstration (total spending)	Deployment (additional cost of climate technologies)		Diffusion (additional cost of climate technologies)		Total
	Global	Global	Global	Developing countries	Global	Developing countries	Global
Current total	15.8–70		30–45	n.a.	31.5–49	11.3–18.8	77.3–164
Additional financing needed	50[a] 20–100[b] 10[c] 30–100[d]	27–36[e]	100–200[f] 57–94[g] 25–35[h]	10–120[i]	1100[j] 379.5[k] 317–811[l]	130–660[m]	379–1436

Notes: n.a. = not available.

[a] Stern et al (2006, p371); public finance only.

[b] Doornbosch et al (2008, p5).

[c] UNFCCC (2007, p7); public finance only.

[d] Nemet and Kammen (2007, Table 1, p752); US government energy R&D spending only.

[e] Calculated from demonstration costs estimated in IEA (2008b, Chapter 3).

[f] Doornbosch, et al (2008); estimates assume a global carbon price of US$25 tonnes of carbon dioxide equivalent.

[g] UNFCCC (2007, p90), based on Stern (2006) and IEA (2008b) estimates of existing deployment support.

[h] UNFCCC (2007, p6).

[i] IEA (2008a) estimates that about 60 per cent of investment is needed in developing countries.

[j] IEA (2008b, p39 and Chapter 6).

[k] UNFCCC (2007, Table IX-64, p175).

[l] McKinsey (2009); lower figure is for 2011–2015 and higher figure is for 2026–2030.

[m] For the low end of the range, the level of investment required in developing countries is calculated using the same investment share as estimated by the UNFCCC, which is 40.9 per cent in developing countries and 59.1 per cent in developed countries (UNFCCC, 2007, Table 4, Annex V, p214). For the higher end of the range the investment share is 60 per cent for developing countries and 40 per cent for developed countries as estimated by IEA (2008a, p240).

Source: UNFCCC (2009)

public policies and investments can leverage the private sector. Figure 24.4 summarizes the private-sector leveraging potential of a wide range of public policy and financing options. On the vertical axis public policy and investment instruments are ranked according to their ability to leverage additional public and private sector financing.

As previously discussed, public financing for climate change may be restricted either because the new climate change agreement is unable to secure a full commitment from Annex II countries to provide the necessary financing for climate change, or because economic circumstances or the competing priorities for public financing do not allow for sufficient public financing to be provided.

In Table 24.3 the public and private shares of finance are estimated under three scenarios using the high end of the range for required total additional financing. In scenario 1, an estimate is made based on existing leveraging ratios[3] that are currently common across the stages of technological maturity. In scenario 2, the public and private shares are estimated assuming that new policies and measures are adopted and a moderately enhanced leveraging ratio takes affect under the post-2012 climate change agreement. Finally, in scenario 3 a high performance mix of policy instruments is tested in which the leveraging ratios assumed are at the top of the range of what could be possible

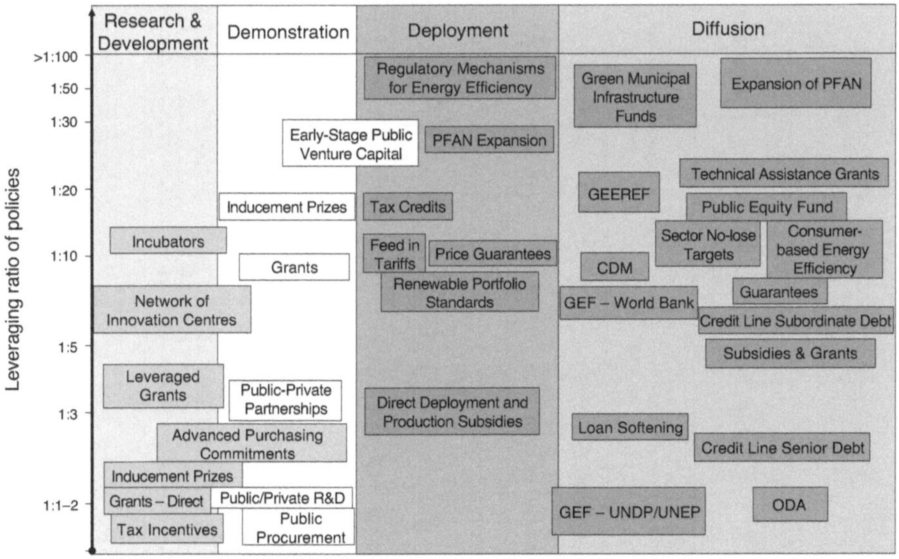

Source: Higham (2009)

Figure 24.4 *Leveraging potential of public policy and financing options for climate change*

Table 24.3 *Three leveraging ratio scenarios for public and private investment shares*

	US$ billion	*Total*	*R&D*	*Demonstration*	*Deployment*		*Diffusion and commercial*	
			Global	*Global*	*Global*	*Developing*	*Global*	*Developing*
	Total annual additional finance	1247–1436	20–100	27–36	100–200	60–120	1100	660
Scenario 1	Estimated existing leverage ratio		1:0.5	1:1	1:4	1:2	1:10	1:10
	Private	1086–1191	7–33	14–18	75–150	40–90	990	594
	Public	162–245	13–67	14–18	25–50	20–30	110	66
Scenario 2	Moderately enhanced leverage ratio		1:1	1:3	1:5	1:5	1:10	1:10
	Private	1155–1227	10–50	20–27	80–160	50–100	990	594
	Public	92–379	10–50	7–9	20–40	10–20	110	66
Scenario 3	High performance-enhanced leverage ratio		1:2	1:5	1:10	1:10	1:20	1:20
	Private	1853–2006	15–75	22–2[illegible]	90–180	54–108	1045	627
	Public	114–152	5–25	5–7	10–20	6–12	55	33

Source: Based on data in UNFCCC, 2009

in the post-2012 environment. Even under the most optimistic scenario for leveraging private-sector investment, an additional US$114 billion to $152 billion per annum would need to be made available by the public sector for investment into climate change technologies.

The reverse can also be tested by asking what leveraging ratios would be necessary if public investment is limited to perhaps US$50 billion or US$100 billion per annum. In order to test this scenario it is assumed that actual investments are roughly proportional to the investment needs across the various stages of technological maturity. In scenario 1 it is assumed that total additional public investment in research and development increases by US$5 billion per annum, demonstration investments by US$5 billion per annum, deployment investments by US$15 billion and diffusion investments by US$25 billion. In scenario 2, where US$100 billion are allocated per annum, US$10 billion per annum is allocated for research and development, US$10 billion per annum for demonstration, US$30 billion for deployment and US$50 billion for diffusion. The results are presented in Table 24.4.

While crude, under these scenarios it may be achievable to meet financing needs for climate change if public financing is limited to an additional US$100 billion per annum; however, it appears to be beyond the capability of prospective policies and measures to leverage the private sector sufficiently if only US$50 billion per annum of public investment was available. In the later case, leveraging ratios for diffusion of technology would need to increase to about 1:40, which could mean much greater levels of regulatory intervention than currently envisaged, or politically feasible.

It might be feasible to set and work towards a target for enhancing the leveraging ratio of the convention. By making larger public endowments to be invested in climate change technologies in the early years, the expectation is that as public policies become more refined and effective through the process of learning by doing, the public investment can be reduced and private investment can take a stronger role.

It is yet to be seen how the policies surveyed may be able to be scaled up, as this would have a significant bearing on where at the international level the greatest effort should be made. Ideally, total investment scale of each policy would be estimated and together with leverage ratios and other factors in mind, a more useful assessment of financial needs could be made. It is clear that a mix of policy instruments is required for international climate change policy, but the exact mix in different regions is not well known.

The research undertaken in support of the post-2012 climate change negotiations has estimated the scale of financing needed, where that investment is needed, what innovative financing options are available, and how effective public policies are and could be in leveraging the private sector. But what policy options are available, and how can they be combined to form a coherent strategy for a low-carbon, climate change-ready world?

Table 24.4 *Implications of restricted public funding for leveraging ratios if total investment continues to meet IEA (2008b) estimates of additional financing needs for technology to 2050*

	US$ billion	*Total*	*R&D*	*Demonstration*	*Deployment*		*Diffusion and commercial*	
			Global	*Global*	*Global*	*Developing*	*Global*	*Developing*
	Total annual additional finance	1247–1436	20–100	27–36	100–200	60–120	1100	660
Scenario 1 – US50 billion public investment per annum	Required leverage ratio		1:3–19	1:5–6	1:6–12	1:6–12	1:43	1:43
	Private	1197–1386	15–95	22–31	85–185	51–111	1075	645
	Public	50	5	5	15	9	25	15
Scenario 2 – US100 billion public investment per annum	Required leverage ratio		1:1–9	1:2–3	1:3–6	1:3–6	1:21	1:21
	Private	1147–1336	10–90	17–26	70–170	42–102	1050	630
	Public	100	10	10	30	18	50	30

Source: Based on data in UNFCCC, 2009

Policy options for financing climate change technologies

Financing options for climate change technologies vary according to specific technology and country circumstances. Different financial vehicles target technologies in each stage of technological maturity. Typically, a complex mix of types of finance and financial instruments must be used simultaneously. It is important to tailor financing policy options to closely match the types of finance and financing instruments needed.

There are many different types of financing that can be applied to a climate change technology project or programme of activities; however, in general, there are three main types of finance: equity, mezzanine and debt finance. In practice, firms often acquire much more complex financing arrangements than this general typology of finance suggests.

Figure 24.5 illustrates the characteristics of each major type of finance.

Public financing mechanisms can be tailored to address each type of finance. Measures that aim to support early stage technology development tend to be equity-based financial instruments; mezzanine financing is particularly suited to assist in the commercialization of technologies; and debt-financing measures are often applied to low-risk technologies and are suited to deployment and diffusion of technologies. An example of the latter is the World Bank's Climate Change Bond, which was issued in November 2008 and raised US$344 million for investment in climate change mitigation projects in developing countries. These types of initiatives, while small in comparison to financial needs, are particularly important in the context of the state of the financial institutions and markets in 2008 to 2009 and the need to ensure growth in climate change investments.

A company will typically attempt to finance its operations with a maximum of debt financing because this will allow the company to minimize the cost of finance, and to maximize its ongoing control of the company and the amount

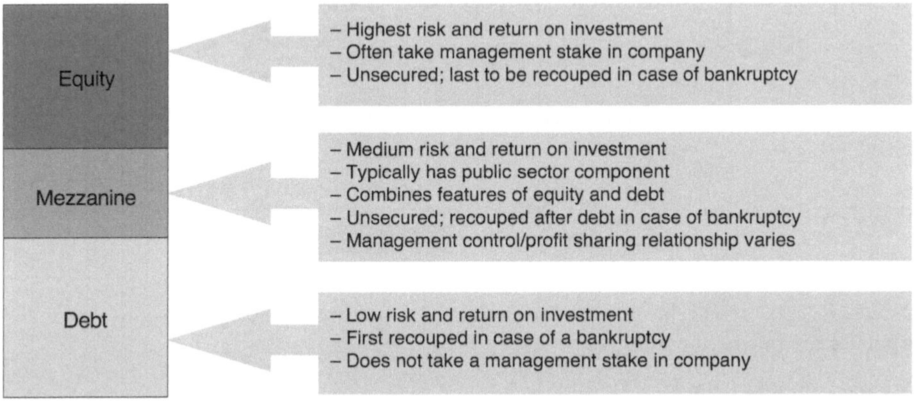

Figure 24.5 *Characteristics of the major types of finance*

of revenue that it will retain for future development of the company. However, debt financing may be difficult to obtain due to the inherent risks involved in new technologies or new companies.

Figure 24.6 shows the type of financing instruments required by stage of technological maturity. Companies in the early stages of development are heavily dependent upon government grants, angel investors and the personal funds of the technology developer. As a company takes its technology closer to commercialization, venture capital and private equity financing become more common, and then in the commercial stage, debt financing tends to dominate.

The level of early stage venture capital has not grown significantly since 2001. Analysis by UNEP (2008) has demonstrated the need for public-sector venture capital financing tools to fill these gaps due to the reluctance of the private sector to bear the significant risks associated with early stage financing.

The vast majority of early stage venture capital is the public investment needed to stimulate later stage private investments, suggesting that a global

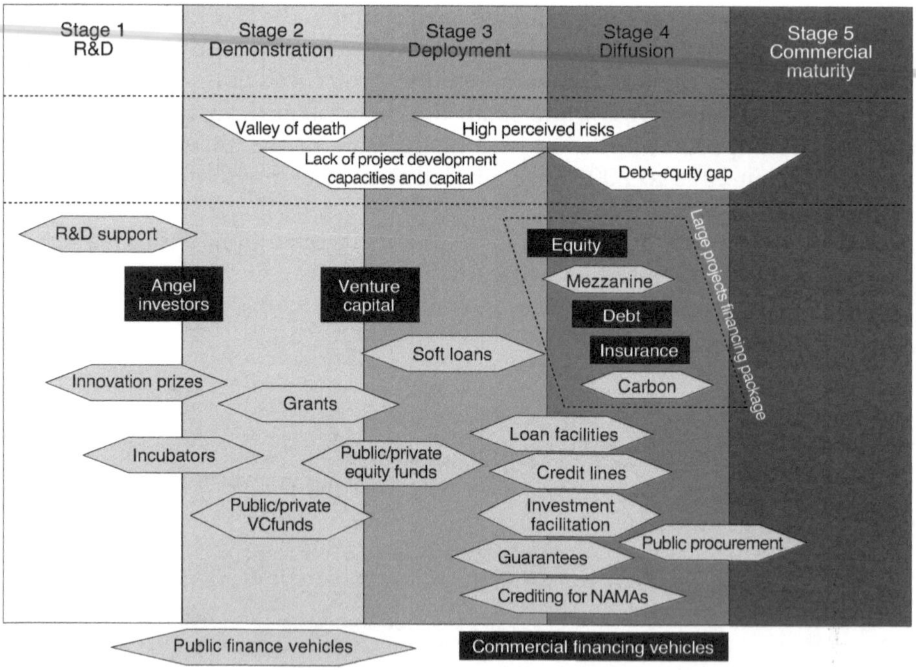

Notes: R&D = research and development; VC = venture capital; NAMAs = nationally appropriate mitigation actions.

Source: adapted from UNEP (2008)

Figure 24.6 *Financing vehicles by stage of technological maturity*

public venture capital fund capitalized with US$28.6 billion to $34.8 billion is required to help leverage subsequent venture capital and private equity investments to bring new technologies into deployment.

The International Energy Agency (IEA, 2008b) and the UNFCCC (2007, 2008b, 2009) have estimated that an additional US$100 billion to $120 billion per annum in early deployment support for technologies in developing countries will be necessary as part of an overall financing strategy to meet a 500 to 550 parts per million carbon dioxide equivalent stabilization level.

Under the convention these are additional or incremental costs that are to be financed by developed countries (Annex II parties). However, the types of mechanisms that can most efficiently drive the early deployment of technologies are national market-based mechanisms which subsidize these more expensive technologies and make them attractive relative to the cost of incumbent technologies that are more emissions intensive. Project-based approaches such as the CDM have not been designed to support early stage technologies, and project-based mechanisms such as the Global Environment Facility (GEF) would need to be scaled up by an order of magnitude and may be unwieldy. For these reasons, a policy-based approach is more desirable.

A policy-based approach would support developing countries to finance national policies such as renewable energy targets through policy instruments such as feed-in tariffs or renewable energy obligations for electricity generators. With technical assistance from developed countries, the national government would develop strategies for implementing national policies to drive early deployment of climate change technologies. A financial mechanism under the convention would provide the appropriate financing package to support the implementation of the policy. This may involve a combination of financing instruments, including direct grants, concessional loans, carbon crediting and other forms of support. The financing may be conditional upon national policy reforms and drive the necessary improvements to the 'enabling environments' of developing countries (Metz et al, 2000).

Another option for early deployment of technology may be to allow for a portion of a developed country's renewable energy obligations to be fulfilled from new renewable energy projects in developing countries (UNFCCC, 2007). This would have the benefit of reducing the overall cost of renewable energy obligations in developed countries (IEA, 2005) and would provide an additional source of financing for early deployment of technologies in developing countries.

Parties to the convention have suggested a wide range of options to create new financing sources and vehicles to enhance technology development and transfer. The processes that have been instigated through the Bali Action Plan have also stimulated the development of many proposals from international organizations and experts.

Only the proposals relating to technology development and transfer are considered here. Table 24.5 contains a list of the proposals and options relating to technology development and transfer submitted by parties or organizations. Elements of proposals with strong similarities are combined. The proposals and options have been classified by the stage of technological maturity they address.

Many of the options have strengths that make them suitable for specific purposes. No single option is intended to address all of the financing gaps and barriers across all sectors and stages of technological maturity. Rather, enhanced financing for technology development and transfer will consist of a package of options. In constructing such a package, complementarity between the options needs to be kept in mind. It is also important to include sunset clauses and exit strategies for many financing policies because, over time, it is expected that most climate change mitigation technologies will became fully commercial and will no longer require public financial support.

The role of the Copenhagen Climate Change Treaty

The convention and the Kyoto Protocol foresee financial assistance from developed country parties to developing country parties. Developed country parties (Annex II parties) committed to provide new and additional financial resources to assist developing country parties comply with their obligations under the convention (Article 4.3) and the Kyoto Protocol (Article 11.2). The financial assistance may be provided through a 'financial mechanism' established by Article 11 of the convention or through bilateral, regional or other multilateral channels.

The GEF was designated as an entity entrusted with the operation of the financial mechanism of the convention on an interim basis in 1995. The financial mechanism is accountable to the COP, which decides on its policies, programme priorities and funding criteria. However, it is a very minor player in the financing of climate change – representing less than 1 per cent of total investment.

The fourth review of the financial mechanism has been initiated and will be completed by the COP at its 15th session in 2009. The review will determine the fifth replenishment of the GEF. A key question that should be addressed through the current review of the financial mechanism is what role the financial mechanism should play in the overall financing challenge for climate change. Should it aim to take a niche role in financing by targeting key gaps, or should it also take a strategic role in financing for climate change across all sources and technology stages?

While the proliferation of financing mechanisms outside of the convention is an encouraging sign in the sense that it creates competition and innovation as each entity finds a suitable role within the financing landscape, it has also resulted in duplication and lack of overall efficiency. It is a sign that there is a lack of leadership on the issue of financing climate change as each institution is vying for power to take the commanding role. The convention is the pre-eminent expression of global efforts to address climate change, and surely it should embody the necessary political power that can provide the leadership and direction for financing, maximize efficiency and shape the financing landscape so that it delivers in a manner consistent with the objectives of the convention.

Table 24.5 *Options for enhancing convention finance for technology*

Maturity stage	*Proposal*	*Options*
Research, development and demonstration	Targets for national research, development and demonstration expenditure	• Targets for the provision of financial support for research and development in developing countries • Targets for reducing or eliminating support for research, development and demonstration for environmentally harmful technologies
	Convention Research, Development and Demonstration Fund	• Pooling of national research, development and demonstration expenditures • Developing countries' financial assistance for participation in international technology agreements • Investment guarantee or risk reduction tool
	Global Network of Innovation Centres	• Public–private partnerships • Intellectual property sharing • Technology transfer • Various financing tools
	National targets for technology demonstration	• Global technology roadmaps • Technology agreements • Commitments to demonstration financing in developing countries
	Innovation prizes	
	UNFCCC Technology Fund	Could include the following sub-funds: • renewable energy; • venture capital; • public equity; • mezzanine finance; • investment risk tools
Deployment	National targets for technology deployment	• Five- to ten-year national targets • Financial support through the financial mechanism of the convention or the proposed technology fund
	Public procurement mechanism	• Tendering programme • Price guarantees • Coordinated public procurement • Advanced purchasing commitments
	International project development mechanism	• Market analysis • Programme/large-scale project feasibility and scoping • Structure financial packages
Diffusion	Carbon financing	• Expansion of interlinked domestic emissions trading schemes • Enhanced/expanded project-based Clean Development Mechanism (CDM) • Expansion of the CDM through scale-up of programmatic approaches • Sectoral approach to the CDM • Sector no-lose targets and crediting • Crediting for nationally appropriate mitigation actions

Table 24.5 *Options for enhancing convention finance for technology*

Maturity stage	*Proposal*	*Options*
Diffusion	Technology agreements and programmes	• Sectoral technology-oriented agreements (priority for steel production, coal-fired power plants, cement and road transportation) • International technology barriers programme to address barriers faced by specific technologies; this could include purchase of licences or patents • Global adoption of energy-efficiency standards and mandates • Technology scale-up partnerships
	International investment facilitation	• Expansion of the Private Finance Advisory Network and other similar investment facilitation programmes
	Concessional financing	• Loan facility for energy-efficiency measures • Credit line for senior debt; green bonds
	National renewable energy and energy efficiency targets	• Support for commitments to national renewable and energy-efficiency targets in developing countries
	Scale-up Convention Financial Mechanism	• Fifth replenishment of the Global Environment Facility (GEF)
Technology Transfer	National technology transfer plans	• Country driven building upon technology needs assessments and national adaptation plans of action
	Export credit agencies	• Use financial instruments to provide reduced interest rates, credit guarantees and insurances for technology transfer • Limit support for technology exports by their export credit agencies to environmentally sound technologies
	UNEP-proposed International Technology Transfer Programme	• An integrated programme of related initiatives

Source: Haites et al (2009)

This is not to say that the actual mechanisms for financing emissions reductions projects should reside within the convention. Such an approach would essentially require the creation of a new financing institution under the convention with similar capability to the Bretton Woods institutions. It should be possible to harness the existing institutions to deliver the objectives of the convention. However, to do so requires a financing mechanism under the convention that has both the authority and the means to do so. The moral suasion of the UNFCCC is not enough. In terms of the means, adequate financial leverage under the convention is critical. It would be necessary for the convention to possess sufficient financial resources to leverage shifts in global financial institutions on the scale necessary.

The financial mechanism of the convention could function as a large fund of funds. Under the guidance of the COP, it would define the desired global strategic outcomes and how to hold to account the recipient financial institutions, and make the allocation of funds conditional upon reforms in these institutions that further contribute toward the financing of climate change technologies.

A possible model for an enhanced technology financial mechanism of the convention is illustrated in Figure 24.7, with policy options corresponding to those presented in Table 24.5.

Conclusions

The post-2012 climate change agreement cannot hope to leverage private-sector investment on the scale needed unless strong climate change targets are agreed by Annex I countries and commitments by developing countries to reduce emissions below the business-as-usual scenario. These commitments will be the key drivers for the private sector, which would also benefit from a long-term target for emissions reductions.

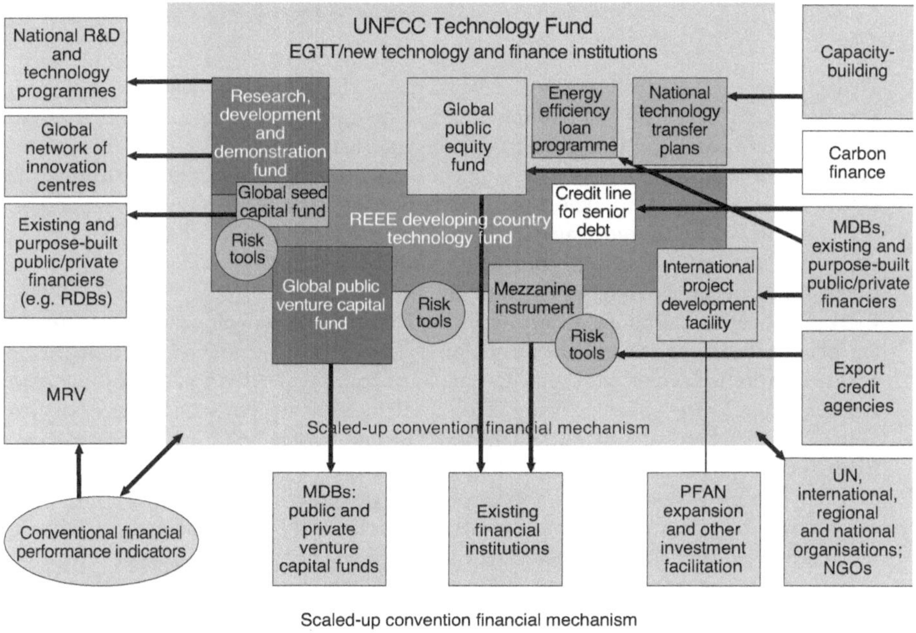

Note: EGTT = Expert Group on Technology Transfer, MDB = multilateral development banks, MRV = measurable, reportable and verifyable, NGOs = non-government organizations, PFAN = private financing advisory network, R&D = research and development, RDBs = regional development banks, REEE = Renewable Energy and Energy Efficiency Fund.

Figure 24.7 *A potential model for an enhanced technology financing mechanism under the convention*

Public financing will be crucial to achieving the objectives of the convention; however, there is a need to overhaul the financial mechanism of the convention and raise its prominence so that it can shape the overall financial landscape as it affects climate change.

It is difficult to escape the need for a large new specified fund for targeted investments in global climate change priorities. The UNFCCC (2008b) has surveyed the current proposals for new sources of finance and found that there are many options which could meet the needs for financing – but they will ultimately depend upon contributions from national budgets. Whether a new fund should be established under the convention, or housed within a Bretton Woods institution or the GEF, is a matter for judgement. However, creating a new fund of funds under the convention has the distinct advantage of creating certainty while maximizing the potential of existing financial institutions. This is crucial given the radical shifts in financing that is required. Research undertaken for the UNFCCC indicates that convention financing in the order of US$50 billion to $100 billion per annum would be ideal. Whether it is possible to negotiate as part of the Copenhagen agreement is to be seen.

Notes

1 This chapter is based on work undertaken for the UNFCCC Expert Group on Technology Transfer (EGTT) supporting the post-2012 negotiations and the Bali (COP 13) mandate, and reflects contributions from many other researchers and organizations with an interest in a successful post-2012 negotiation under the UNFCCC in terms of financing the convention and the technology development, deployment, diffusion and transfer challenges posed by climate change. The views expressed are those of the author and should not be attributed to those of the UNFCCC, Energy Research Centre of the Netherlands (ECN) or the EGTT. This work has been undertaken jointly with Dr Heleen de Coninck and Erik Haites. I am also grateful for the contributions and advice from Xander van Tilburg, Eric Massey, Laurens Bouwer, Zhu Liucai, Balawant Joshi, Jose Alberto Garibaldi, Stefan Bakker, Ester Basri, Silvia de Vaan, Dolf Gielen, Michael Taylor, Raouf Saidi, Jos Sijm, Eric Usher, Joost van Putten, David Vincent, Florin Vladu, Xiaohua Zhang, Morgan Bazilian and other members of the EGTT.

2 The Bali Action Plan envisaged a Copenhagen Agreement with:
- measurable, reportable and verifiable nationally appropriate mitigation commitments or actions, including quantified emission limitation and reduction objectives, by all developed country parties, while ensuring the comparability of efforts among them, taking into account differences in their national circumstances; and
- nationally appropriate mitigation actions by developing country parties in the context of sustainable development, supported and enabled by technology, financing and capacity-building, in a measurable, reportable and verifiable manner.

3 The terms leveraging and mobilization are used interchangeably within this chapter as they are within the literature and within the international climate change negotiations. The leveraging ratios used in this chapter are expressed mathematically as x:y, where x is the amount of public-sector investment, and y is the amount of

private-sector finance that is a direct consequence of the public investment (see Higham, 2009).

References

Auerswald, P. and L. Branscomb (2003) 'Valleys of death and Darwinian seas: Financing the invention to innovation transition in the United States', *Journal of Technology Transfer*, vol 28, pp227–239

Auerswald, P., L. Branscomb, N. Demos and B. Min (2005) *Understanding Private-Sector Decision Making for Early-Stage Technology Development: A Between Invention and Innovation Project Report*, National Institute for Standards and Technology (NIST), US Department of Commerce report, NIST GCR 02-841A, US

Branscomb, L. and P. Auerswald (2002) *Between Invention and Innovation: An Analysis of Funding for Early-Stage Technology Development*, Advanced Technology Program, National Institute for Standards and Technology (NIST), US Department of Commerce Report, NIST GCR 02841, US

Coogan, A., R. Earhart, B. Eggl, V. Handke, C. Kamburow, V. Krauth, A. Marshall and R. van Ermen (2007) *FUNDETEC Study: Comparison and Assessment of Funding Schemes for the Development of New Activities and Investments in Environmental Technologies*, European Commission, Brussels

den Elzen, M. and N. Höhne (2008) 'Reductions of greenhouse gas emissions in Annex I and non-Annex I countries for meeting concentration stabilisation targets: An editorial comment', *Climatic Change*, vol 91, pp249–274

Doornbosch, R., D. Gielen and P. Koutstaal (2008) *Mobilising Investments in Low-Emission Energy Technologies on the Scale Needed to Reduce the Risks of Climate Change*, OECD SG/SD/RT(2008)1, Paris, France

Flåm, K, and J. Skjærseth (2009) 'Does adequate financing exist for adaptation in developing countries?', *Climate Policy*, vol 9, no 1, pp109–114

Haites, E., A. J. Higham and H. De Coninck (2009) *Towards Financing Technology: Recommendations on Future Financing Options and Risk Mitigation Tools Necessary for Enhancing the Implementation of the Convention with Regard to the Technology Transfer Framework*, Background Paper 3 for the Expert Group on Technology Transfer, UNFCCC, Bonn

Halsnæs, K., P. Shukla, D. Ahuja, G. Akumu, R. Beale, J. Edmonds, C. Gollier, A. Grübler, M. Ha Dong, A. Markandya, M. McFarland, E. Nikitina, T. Sugiyama, A. Villavicencio and J. Zou (2007) 'Framing issues', in B. Metz, O. Davidson, P. Bosch, R. Dave and L. Meyer (eds) *Climate Change 2007: Mitigation, Contribution of Working Group III to the Fourth Assessment Report of the Intergovernmental Panel on Climate Change*, Cambridge University Press, Cambridge and New York, NY

Higham, A. J. (2009) *Other People's Money: Enhancing Public Sector Leverage of Private Sector Investment for Climate Change Mitigation*, ECN, Amsterdam

IEA (International Energy Authority) (2005) *Deploying Climate-Friendly Technologies through Collaboration with Developing Countries*, IEA Information Paper, Paris, France

IEA (2007) *World Energy Outlook 2006*, IEA, Paris, France

IEA (2008a) *World Energy Outlook 2008*, IEA, Paris, France

IEA (2008b) *Energy Technology Perspectives: Scenarios and Strategies to 2050: In Support of the G8 Plan of Action*, IEA, Paris, France

McKinsey (2009) *Pathways to a Low Carbon Economy: Version 2.0 of the Global Greenhouse Gas Abatement Cost Curve*, McKinsey and Company, USA

Metz, B. (2008) *Emission Reductions for Stabilisation and Mitigation Potentials: Information from the IPCC Working Group III AR4 and New Studies*, AWG-KP Workshop on Mitigation Potentials, Poznań, Poland, 3 December 2008

Metz, B., O. Davidson, J.-W. Martens, S. van Rooijen and L. van Wie McGory (eds) (2000) *Methodological and Technological Issues in Technology Transfer, Special Report of IPCC Working Group III*, Cambridge University Press, Cambridge and New York, NY

Murphy, L. and P. Edwards (2003) *Bridging the Valley of Death: Transitioning from Public to Private Sector Financing*, NREL/MP-720-34036, NREL, Colorado, US

Nemet, G. and D. Kammen (2007) 'US energy research and development: Declining investment, increasing need, and the feasibility of expansion', *Energy Policy*, vol 35, pp746–755

New Energy Finance (2008) *Global Trends in Sustainable Energy Investment 2008*, New Energy Finance, London

New Energy Finance and UNEP (United Nations Environment Programme) (2008) *Public Venture Capital Study*, SEF Alliance, UNEP, Paris

OECD (Organisation for Economic Co-operation and Development) (1997) *National Innovation Systems*, OECD, Paris, France

Raworth, K. (2007) *Adapting to Climate Change: What's Needed in Poor Countries and Who Should Pay*, Oxfam International, www.oxfam.org/en/files/bp104_climate_change_0705.pdf/download

Stern, N., S. Peters, V. Bakhshi, A. Bowen, C. Cameron, S. Catovsky, D. Crane, S. Cruickshank, S. Dietz, N. Edmonson, S.-L. Garbett, L. Hamid, G. Hoffman, D. Ingram, B. Jones, N. Patmore, H. Radcliffe, R. Sathiyarajah, M. Stock, C. Taylor, T. Vernon, H. Wanjie and D. Zenghelis (2006) *Stern Review: The Economics of Climate Change,* HM Treasury, London.

UNEP (United Nations Environment Programme) (2002) *Barriers to Sustainable Energy Finance*, UNEP-SEFI, Paris, France

UNEP (2007) *Executive Briefing: Making it Happen: Renewable Energy Finance and the Role of Export Credit Agencies*, UNEP-SEFI, Paris, France

UNEP (2008) *Public Finance Instruments for Climate Mitigation: Options Document*, UNEP-SEFI, Paris, France

UNFCCC (United Nations Framework Convention on Climate Change) (2007) *Investment and Financial Flows to Address Climate Change*, UNFCCC, Bonn

UNFCCC (2008a) 'Ideas and proposals on paragraph 1 of the Bali Action Plan', note by the Chair, Adhoc Working group on long term cooperation under the convention, Bonn, Germany

UNFCCC (2008b) 'Identifying, analysing and assessing existing and potential new financing resources and relevant vehicles to support the development, deployment, diffusion and transfer of environmentally sound technologies', interim report by the Chair of the Expert Group on Technology Transfer, Bonn, Germany

UNFCCC (2009) 'Recommendations on future financing options for enhancing the development, deployment, diffusion and transfer of technologies under the Convention', report by the chair of the Expert Group on Technology Transfer to the Adhoc Working Group on Long Term Cooperation under the Convention, Bonn, Germany

UNFCCC (2009) *Existing and Potential New Financing Resources and Relevant Vehicles to Support the Development, Deployment, Diffusion and Transfer of Environmentally Sound Technologies*, Report to the AWG-LCA, UNFCCC, Bonn

US SBA (1994) *Bridging The Valley of Death: Financing Technology for a Sustainable Future*, Report to the US EPA, December 1994, SBA, Washington, DC
Williams, E. (2004) *Crossing the Valley of Death*, Warwick Ventures, University of Warwick, www2.warwick.ac.uk/services/ventures/valley.pdf.

Index